INFORMATION TECHNOLOGY MANAGEMENT

A Knowledge Repository

Jay Liebowitz

CRC Press

Boca Raton Boston London New York Washington, D.C.

Dedication

This book is dedicated to those who believe in knowledge sharing of the highest order and who are willing to admit to failure and learn from those experiences.

To Janet, Jason, Kenny, my parents, and all my students.

Contents

SECTION III: A CODE OF ETHICS

SECTION IV: FUTURE CONSIDERATIONS

Preface

No matter the endeavor, no one is eagerly willing to admit their mistakes. In information technology, the failure of a technology or project is usually management-related. As Lewis Carroll once wrote, "If you don't know where you are going, then any road will lead you there."

In trying to be as concise and succinct as possible, this book presents a compendium of some short cases, anecdotes, and vignettes dealing with the *management* of information technology. The material in this book is mainly based on my personal experiences with a variety of organizations. As my background is mostly in the expert systems field, the book is heavily slanted towards cases involving management relating to expert systems technology. In spite of this emphasis, one can merely replace the term "expert system" with "information system" technology in almost all of the case studies, and many of the lessons would hold true for information systems in general.

This book is geared primarily for managers and developers of information technology. I have always felt that one learns more from failures than from successes. (That is why I also edit an international journal, *Failure and Lessons Learned in Information Technology Management,* published by Cognizant Communication Corporation, Elmsford, NY.) Thus, these bittersweet case studies will hopefully provide you with more insight into the management of information technology by perhaps learning from my mistakes. I also hope that this will be a stimulus for others to cleanse their souls in sharing not only the successful cases but also the less successful ones.

Enjoy the book!

Jay Liebowitz, D.Sc.

Jay's Top Ten List of Information Technology Management Lessons Learned (From Least Important to Most Important)

10. Feel comfortable with risk vs. return.
9. Knowledge management will be a critical success factor in those organizations that will be the leaders of their field.
8. Invest in employee training and R&D for continued company growth.
7. No system is any good if no one wants to use it.
6. Carefully examine the corporate culture and determine how best to deal with "resistance to change."
5. Without the proper implementation and organizational concerns analyzed up front, the information system (IS) could be a technical success but a technology transfer failure.
4. "Solve the business problem" — don't force-fit information technology (IT) to requirements.
3. Find senior-level champions within the organization to help "carry the ball" for financial and moral support of the IS/IT.
2. Show how IT contributes to the "bottom line."
1. *Information technology is an integral part of an organization and must be integrated within the business/strategic plan of the company.*

Information Technology Management: Success or Failure?

Don't Get Lost: Tie the Information Systems Strategy to the Strategic Mission of the Firm

Many information systems (IS) projects fail. According to various studies, only about 20% to 30% of IS projects are successful from project inception through institutionalization. The technology is usually *not* the limiting cause. Rather, *management* of the technology is typically the culprit in the lack of success of an information system.

How can system developers and information technology managers better ensure that an information system will not lead to failure? According to Stephen Flowers, Director of the Center for Management Development at the University of Brighton, Great Britain, there are several critical failure factors associated with information systems:

1. Organizational context:
    ■ Fear-based culture
    ■ Poor reporting structure
2. Management of project:
    ■ Overcommitment
    ■ Political pressures
3. Conduct of project:
    ■ Technology focus
    ■ Leading edge system
    ■ Complexity underestimated
    ■ Technical "fix" sought
    ■ Poor consultation
    ■ Changing requirements
    ■ Weak procurement
    ■ Development sites split
    ■ Project timetable slippage
    ■ Inadequate testing
    ■ Poor training

These three areas of organizational context, the management of the project, and the conduct of the project must be carried out properly in order for an information systems project to succeed.

The Secret to Success — PERMS

The secret to a (hopefully) successful information systems project is expressed by the acronym PERMS (people, expectations, resources, money, and scheduling).

"People" refers to the system developers, users, and managers. The users should be involved in the design, development, implementation, and evaluation processes. Their input and feedback are crucial throughout the system development life cycle. Managers need to back the project, both morally and financially. Having a project champion and top management endorser is an important contributing factor for a successful information system. The systems developers should also be prepared to integrate user and management concerns into the design and development of the information system.

"Expectations" refers to keeping realistic goals and potential accomplishments as related to the information system development and implementation. In the early days of artificial intelligence (AI), there was tremendous hype and enthusiasm, which often resulted in overexpectations as to what the AI system could deliver. The project champion, project leader, management sponsor, and the system development team must be careful to present realistic goals and objectives for their information system. Of course, some marketing is also needed to keep the project in the necessary profile, but again, puffing and undue advertising of the system will lead the project down the road to failure.

"Resources" refers to the necessary human, hardware, and software resources needed to develop, implement, and maintain the information system. Training costs are also a part of this category, as the users and maintainers must be trained in how to operate and update the system, respectively. The necessary equipment and software packages must also be procured to develop the information system.

"Money," which really falls under "resources," is highlighted here to accent the importance of having the necessary financial resources to develop and transition the system within the organization. Studies have indicated that two thirds of the total system life cycle costs stem from the maintenance stage. Thus, there must be adequate financial resources to cover the front-end and back-end of the life cycle.

"Scheduling" is the "S" in PERMS. Scheduling refers to the allocation of resources over time for building and implementing the information system. Proper configuration management control is necessary to ensure that deliverables are delivered on time, within budget, and according to the requirements.

Be Careful!

There are several caveats that an information systems project leader, developer, or manager should be aware of when building an information system. Some of these are highlighted by Tom Beckman, an information systems developer for the U.S. Internal Revenue Service:

- People generally don't like to be measured.
- Executive champions are necessary, but not sufficient. Even champions can't make information systems successful unless user executives are also convinced.
- Never underestimate the power and stupidity of politics and the funding agency (often not the user).
- Technical issues may be the least of your concerns.
- Managers and users want systems that automate tasks in ways that look similar to current methods.
- Be sure that managers from user and funding organizations are intimately involved in the project on a regular basis throughout the system's life cycle.
- User and management expectations must be carefully managed and often played down.
- Address user resistance to change.
- In the initial implementation cycle, automate the task just as it is now done. As inefficient as this is, it is highly advised in order to avoid the deadly client management resistance.
- Once the client management is convinced that the technology works, only then can the designer pursue changes that improve the work process and may possibly change client management policy.
- It is essential to engage managers from the client organization on your project team. Involve client managers on a regular basis for ideas and project review.

KEEP RUNNING, BUT MAKE SURE YOU HAVE YOUR SHOES ON!

1 A Knowledge Repository of Lessons Learned: Their Need and Importance

KNOWLEDGE REPOSITORY

- **Best practices**

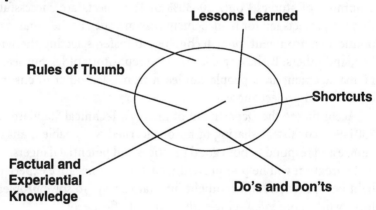

Lessons Learned

Rules of Thumb

Shortcuts

Factual and Experiential Knowledge

Do's and Don'ts

Most managers feel that the critical asset that separates their organization from their competitors is the knowledge assets or intellectual capital of the employees in their organization.[1,2] With many organizations reengineering, downsizing, rightsizing, outsourcing, and the like, the ability to capture, share, and apply the "lessons learned" of the employees (especially those experts who retire or leave the firm) is critical to the success and growth of an organization.

According to the Department of Energy Lessons Learned Office, a "lesson learned" is a good work practice or innovative approach that is captured and shared to promote repeat application. A lesson learned may also be an adverse work practice or experience that is captured and shared to avoid recurrence.

The National Aeronautics and Space Administration's (NASA) definition of a lesson learned is "knowledge or understanding gained by experience. The experience may be positive, as in a successful test or mission, or negative, as in a mishap or failure. Successes are also considered sources of lessons learned. A lesson must be significant in that it has a real or assumed impact on operations; valid in that it is factually and technically correct; and applicable in that it identifies a specific design, process, or decision that reduces or eliminates the potential for failures and mishaps, or reinforces a positive result." NASA has a Lessons Learned Information System, and many organizations in the military, government, and industry are providing a means for knowledge management through their organization. A sample of lessons learned programs is shown in Table 1.

In the information technology management area, there is a great need for developing and using lessons learned knowledge repositories. According to a number of studies,[3] only 20–30% of IS projects are successful. Many of these IS projects fail for management reasons, such as the organizational and political environment, poor technology transfer, selecting the wrong problem, and others. By having a knowledge repository of lessons learned in the IT management area, people can learn from each other to ensure a greater chance of IS project success.

According to the Department of Energy's Technical Standard DOE-STD-7501-95, "consistent sharing of lessons learned will enable managers to recognize and respond to both good practices and potential dangers. A complex-wide program will help to prevent recurrence of negative experiences, highlight best practices, and spotlight innovative ways to solve problems or perform work more safely, efficiently, and cost effectively."

How can these lessons learned be structured? Several organizations have used different lessons learned templates. Following are some examples.

DOE Lessons Learned Template:

   Title: Title of the lesson learned.
   Identifier: Unique identification number to assist in referencing a lesson learned that includes calendar year, operations office identifier, organization or field/area office/contractor identifier, and a sequential number.

Table 1   Sample of Lessons Learned Programs

- Agency for Toxic Substances and Disease Registry
- Australian Environmental Resource Information Network
- Consumer Product Safety Commission
- Department Standards Committee
- Environmental Protection Agency
- FinanceNet
- International Atomic Energy Agency World Atom
- NASA Lessons Learned Information System
- NASA Selected Current Aerospace Notices
- OSHA Standard Interpretations
- U.S. Geological Survey Natural Hazards Programs: Lessons Learned for Reducing Risk
- Air Force LINK Factsheets (Air Force Lessons Learned)
- Automated Lessons Learned Collection and Response System
- Center for Army Lessons Learned (CALL)
- Federal Technology Transfer
- Naval Facilities Engineering Command (Lessons Learned Home Page)
- Product Acquisition and Engineering (Navy Lessons Learned)
- The Code Authority — Underwriters Laboratories Newsletter
- *Failure and Lessons Learned in Information Technology Management: An International Journal* (Cognizant Communication Corp., Elmsford, NY)
- Internet Disaster Information Network
- EnviroWebs Internet Listing of Environmental Information Services
- Department of Energy Lessons Learned Program
- Electric Power Research Institute
- Federal Emergency Management's Global Emergency Management System
- Mayo Clinic and IVI Publishing Inc. Online Health Network
- Total Quality Management Resource Center Lobby

Date: Date the lesson learned was issued.

Originator: Name of the originating organization or contractor.

Contact: Name and phone number of individual to contact for additional information.

Name of Authorized Derivative Classifier: Name of the authorized derivative classifier who determined that the lesson learned document did not contain classified information.

Name of Reviewing Official

Priority Descriptor: A descriptive code that assigns a level of significance to the lesson.

Functional Category(s): The category(s) that best describes the area(s) in which the lesson is most applicable.

Keywords: Words used to convey related concepts or topics stated in the lesson.

References

Lesson Learned Statement: Statement that summarizes the lesson(s) learned from the activity.

Discussion of Activities: Brief description of the facts that resulted in the initiation of the lesson learned.

Analysis: Results of any analysis that was performed, if available.

Recommended Actions: A brief description of management approved actions which were taken, or will be taken, in association with the lesson learned.

According to the Naval Facilities Engineering Command (NAVFAC) (http://www.navy.mil/homepages/navfac/pe/15home.htm), their Lessons Learned Program follows these ground rules:

1. It's an open forum for information relating to lessons learned from research and development, facility design, contracting, construction, maintenance, operation, and related support systems.
2. Successful lessons learned entries will focus on success and solutions to problems generated by field personnel.
3. Field lesson learned coordinators (LLC) and the Naval Facilities Engineering Service Center (NFESC) will act as facilitators to distribute information and to maintain the program.
4. The forum is open to all parties interested in NAVFAC processes and Department of Navy facilities, both public and private.
5. Lessons learned origins will be anonymous. The submitting lessons learned coordinator is listed as the point of contact.

NAVFAC's flow for generating lessons learned is as follows:

1. Field users
   - Originate lessons learned
   - Write synopses on HTML template provided on-line, including key words and the point of contact name, address, and phone number
   - Submit to respective field lessons learned coordinator

2. Field lessons learned coordinator
   - Receive submission of proposed lesson learned from field user
   - Evaluate the proposed lessons learned entry for technical clarity
   - Dispose of proposed lessons learned entry
   - Direct to criteria preparer, when appropriate
   - Coordinate "no action" or local action decision with submitter
   - Prepare lessons learned listing, applying category and key word definitions
   - Check format of listing
   - Obtain security and Public Affairs Office approval
   - Submit to NFESC electronically
3. NFESC
   - Receive submission from field lessons learned coordinator
   - Check and correct format
   - Submit to NFESC operator for inclusion on the WWW home page

The goal of this book is to provide the author's personal experiences and lessons learned in information technology management. Senior management commitment is essential for ensuring that a useful lessons learned program is developed, implemented, supported, and administered. Management is responsible for defining, documenting, and effectively assigning and communicating the roles and responsibilities for the lessons learned program(s) within their organization. According to the Department of Energy Lessons Learned Program, management has the following responsibilities:

- Effective review and analysis of operating, technical, and business experience information;
- Adoption of good work practices;
- Incorporation of appropriate actions in a timely manner;
- Commitment of sufficient resources to effectively implement a lessons learned program (including budgeting, knowledge sharing incentives, and management processes to support incorporation of lessons that are learned into daily activities).

According to DOE and NASA, lessons learned information should be collected and stored in a manner that allows users to identify applicable lessons learned through information searches. Lessons learned programs should include techniques to periodically measure program effectiveness. Stored lessons learned information should be periodically reviewed for usefulness.

Information that is no longer pertinent to organizational activities could be eliminated or archived in accordance with organizational policies and procedures.

Since information and knowledge technology will be a key determinant in the success of an organization over the competitors, having a way to create, store, manage, retrieve, disseminate, and apply lessons learned relating to IT and its management is greatly needed in organizations. A lessons learned program, especially in the application and management of IT, would benefit an organization so that the wheel isn't reinvented and employees and management can learn from the successes and failures of others.

This book, and the following vignettes and lessons learned in IT management, serves as a beginning toward meeting this goal.

References

1. Liebowitz, J. and L. Wilcox (Eds.) (1997), *Knowledge Management and Its Integrative Elements*, CRC Press, Boca Raton, FL.
2. Liebowitz, J. and T. Beckman (1998), *Knowledge Organizations: What Every Manager Should Know*, St. Lucie/CRC Press, Boca Raton, FL.
3. Liebowitz, J. (Ed.) (1997), *Failure and Lessons Learned in Information Technology Management: An International Journal*, Cognizant Communication Corp., Elmsford, NY, Vol. 1, No. 1.

Examples of Actual Lessons Learned Templates from the Web

Factor Met	Evaluation Factor
	Criteria and Guidelines
	Is the lesson relevant?
	Does it have significant bearing within MO&DSD and/or other parts of NASA?
	Is it useful for space-related applications?
	Is the lesson understandable?
	Is it understandable by someone not intimately familiar with the project?
	Is it understandable by someone skilled in the art?
	Have all acronyms been defined?
	Is the level of detail appropriate?
	Is the lesson beneficial?
	Does it have recognizable impact within MO&DSD and/or other parts of NASA?
	Is there a reasonable possibility that a designer or manager could face the same situation or repeat the same mistake?
	Is it a positive lesson concerning an innovative technique or new design that can save money or labor-hours or improve supportability, reliability, or quality?
	Is the lesson valid?
	Is it factually and logically correct?
	Is it consistent with existing laws, regulations, and policies?
	Is it well grounded or justifiable by source?
	Is there evidence that someone will not experience the same problem or that they will get the same outcome if they follow the recommendation?
	Has the approval authority approved the lesson for release?
	Does the approval authority concur with the use of the information in the database?
	Does the approval authority understand the audience for distribution?
	Has the copyright owner concurred with the use of the information in the database?
	Is the lesson realizable?
	Can users implement the recommendation?
	Is it not hypothetical?
	Lesson Components
	Title
	Is it a brief (less than one line), accurate, concise description of the lesson?
	Have the unique features of the lesson been included?
	Does it contain both problem and solution?
	If applicable, does it contain the document section number?
	Description
	Does it contain a concise description of the situation related to a project, procedure, or design?
	If a problem, does it describe what went wrong?
	Has the impact been described?
	Is the environment surrounding the situation adequately described (e.g., related activities, timeframe, guidelines, standards, directives)?
	If a success, has the unique approach, procedure, standard, tool, or technology used been described?
	Is it one or two paragraphs, 100 to 500 characters in length?
	Has additional information, if available, been included?

Figure 1   List of Indicators from NASA Goddard
http://hope.gsfc.nasa.gov/RECALL

Factor Met	Evaluation Factor
	Questions
	Have about six questions been generated or selected?
	Are there some context questions and some confirmation questions?
	For number questions, has the possible range of numbers been specified?
	For list questions, are there no more than six answers?
	If the question needs explanation, has it been provided?
	Action Title
	Is it one line or less?
	Does it clearly and concisely state the single most important finding?
	Does it show a cause-and-effect relationship?
	Action Description
	Does it describe the action a user should take?
	Does it describe who should take the action (e.g., the program manager)?
	Does it describe when a user should take the action (e.g., during what phase)?
	Has additional information, if available, been included?
	If taken from a hardcopy document, has an electronic copy of the complete source document been provided?
	Have other source references been provided, such as a document, presentation, or person with more information?
	Have methods used to achieve success been included?
	Have obstacles to success been included?
	If a problem, has how to avoid the problem in the future been described?
	If a success, has how to ensure that users repeat the success been described?
	If a failure, has how the failure was repaired been described?
	Has the outcome or what happened during the repair been described? Did it turn out as expected?
	Has evidence been included that the recommended repair actually worked (e.g., periodic inspections showed the problem did not recur)?
	Have unsuccessful solutions that were tried and discarded been described?
	General Guidelines
	Has all the requested administrative data been included, such as name, organization, company, address, telephone and fax numbers, project, and dates?
	Can someone else learn from this experience?
	Does the lesson describe a single event?
	Does the lesson describe only one thing that was learned?
	Have differences of opinion been described as separate actions within the same lesson?
	Are acronyms spelled out when they first occur in each lesson part?
	Is terminology consistent throughout the lesson?
	Is terminology understandable by others not familiar with the project?
	Have definitions of terms not commonly understood been provided for the professional language dictionary?
	Have synonyms been provided for multiple words for the same thing in a lesson?

Figure 1 (continued)   List of Indicators from NASA Goddard
http://hope.gsfc.nasa.gov/RECALL

Lessons Learned Template

Title: _____ Identifier: _____

Date: _____ Originator: _____ Contact: _____

Name of Authorized Derivative Classifier: _____

Name of Reviewing Official: _____

Priority Descriptor: _____
Functional Category(s): _____
Keyword(s): _____

References: _____

Lesson Learned Statement: _____

Discussion of Activities: _____
_____
_____
_____
_____

Analysis: _____
_____

Recommended Actions: _____
_____

Figure 2   Lessons Learned Template from Dept. of Energy
http://www.doe.gov/html/techstds/standard/std7501

NAVAL FACILITIES LESSONS LEARNED

Input Form

Basic ground rules:

- Your activity name and your name will NOT appear in the lessons learned database, but we need to have your name and address with this input so we can contact you while we complete the entry for the database.
- Lessons learned must report solved problems, processes or products which avoided problems, or successful process innovations.
- Suggestions for improving criteria may use this form; they will be delivered to the appropriate criteria office.

What is your name?	
What is the name of your Command?	
What is your address?	
Address line 2	
City	
State	
Zip Code	
What is your commercial phone number?	
What is your DSN prefix? (if applicable)	
What is your e-mail address?	
Todays Date	

Figure 3   Naval Facilities Lessons Learned Template
http://ll.nfesc.navy.mil/private/cjodiorne.htm

Topic Area: | Management

Title: (Your descriptive phrase, which should make it easy for a third party to visualize your subject)

Problem Addressed: (A statement of the problem solved. Normally, a very short statement

Description of Initiative or Lesson-Learned:

(A statement of an initiative with results OR a statement of the solution to a problem encountered)

When this form is entered into the lessons learned database, we will catalog it so others can retrieve it based on various search criteria. Most of these catalog selections will be prepared by the Lessons Learned Coordinators. But please help by providing the information listed below:

Type of Facility (if applicable):	
Through what perspective is this lesson learned created and submitted?	Programmer/Planner

Figure 3 (continued)   Naval Facilities Lessons Learned Template
http://ll.nfesc.navy.mil/private/cjodiorne.htm

Discipline
Phase During Which to Implement Solution
Type of Facility
Type of Space
System or Assembly in a Facility
Construction Product or Activity
Climate
Functional Issue
Acquisition or Contracting Strategy
Enabling Tools

Please submit this lessons learned entry to the local Lessons Learned Coordinator serving your activity or the location of the relevant facility, listed below:

After you complete this form, you will be returned to the **Lessons Learned Home Page**.

Figure 3 (continued) Naval Facilities Lessons Learned Template
http://ll.nfesc.navy.mil/private/cjodiorne.htm

LBL Lessons Learned
Input Form

Mail to:
Lessons Learned Coordinator, MS 90-2148

Date: _____

Originator: _____ Phone: _____ FAX: _____

Summary of Event or Finding:

Analysis/Recommendation:

Proposed Interest Areas for Distribution:

Contact for Additional Information

_____

▪ Return to Beginning of Chapter 14, Lessons Learned

▪ Go to Chapter 14 Table of Contents

▪ Go to PUB-3000 Table of Contents

_____

Figure 4 Occupational Safety Group (DOE) Lessons Learned Template
http://ehs.lbl.gov/ehsdiv/pub3000/CH14

ACC UNCLASSIFIED
Lesson Learned Observation Form

Use this form for input of Lessons Learned Only... Any Summary After Action Reports should be input using the Summary - After Action Report (AAR) Format

---

What phase of the OPERATION and/or EXERCISE does this lesson deal with?

```
Readiness
Deploy
Employ
Sustain
Redeploy
```

What is the title of this lesson learned? (Unclassified summary of the observation; 75 characters max)

Title: | (U)

Classification: | (None)
UNCLASSIFIED

Distribution Limit | (None) | (None)

Classification Markings Derived from :

Operation/Exercise Name: (Nickname)

DATE:

Submitted by: Enter your organization :

Enter rank/title & name (optional) **POC:**

Enter your DSN:

Commercial:

5. (?) What is your Observation?: HELP

Figure 5   Lessons Learned Template from the U.S. Air Force
http://redwood.do.langley.af.mil/ADD_Lesson.htm

(U)

6. (?) What is your Discussion?: HELP

(U)

7. (?) What is your Lesson Learned?: HELP

(U)

8. (?) What is your Recommendation?:

(U)

9. (?) What are your Comments?:

(U)

Tell us how to get in touch with you:

Name

E-mail

Figure 5 (continued)   Lessons Learned Template from the U.S. Air Force
http://redwood.do.langley.af.mil/ADD_Lesson.htm

Tel [   ]

FAX [   ]

NOTE: The lesson input will not be published until it goes through the ACC review and validation process.

This form developed by Mr. Randy Schmidt, HQ ACC/DOXE<Betac>.
Revised: Nov 1, 1996

Figure 5 (continued)   Lessons Learned Template from the U.S. Air Force http://redwood.do.langley.af.mil/ADD_Lesson.htm

One Hundred Rules for NASA Project Managers

Lessons Learned as Compiled by Jerry Madden , Associate Director of the Flight Projects Directorate at NASA's Goddard Space Flight Center: (Jerry collected these gems of wisdom over a number of years from various unidentifiable sources. They have been edited by Rod Stewart of Mobile Data Services in Huntsville, Alabama.). January 1, 1995. Updated July 9, 1996.

Table Of Contents

The Project Manager

Rule #1: A project manager should visit everyone who is building anything for his project at least once, should know all the managers on his project (both government and contractor), and know the integration team members. People like to know that the project manager is interested in their work and the best proof is for the manager to visit them and see first hand what they are doing.

Rule #2: A project manager must know what motivates the project contractors (i.e., their award system, their fiscal system, their policies, and their company culture).

Rule #3: Management principles still are the same. It is just that the tools have changed. You still find the right people to do the work and get out of the way so they can do it.

Rule #4: Whoever you deal with, deal fairly. Space is not a big playing field. You may be surprised how often you have to work with the same people. Better they respect you than carry a grudge.

Rule #5: Vicious, dispicable, or thoroughly disliked persons, gentlemen, and ladies can be project managers. Lost souls, procrastinators, and wishywashies can not.

Rule #6: A comfortable project manager is one waiting for his next assignment or one on the verge of failure. Security is not normal to project management.

Rule #7: One problem new managers face is that everyone wants to solve their problems. Old managers were told by senior management—"solve your own darn problems, that is what we hired you to do."

Rule #8: Running fast does not take the place of thinking for yourself. You must take time to smell the roses. For your work, you must take time to understand the consequences of your actions.

Rule #9: The boss may not know how to do the work but he has to know what he wants. The boss had better find out what he expects and wants if he doesn't know. A blind leader tends to go in circles.

Rule #10: Not all successful managers are competent and not all failed managers are incompetent. Luck still plays a part in success or failure but luck favors the competent hard working manager.

Rule #11: Never try to get even for some slight by anyone on the project. It is not good form and it puts you on the same level as the other person and, besides, probably ends up hurting the project getting done.

Rule #12: Don't get too egotistical so that you can't change your position, especially if your personnel tell you that you are wrong. You should cultivate an attitude on the project where your personnel know they can tell you of wrong decisions.

Rule #13: A manager who is his own systems engineer or financial manager is one who will probably try to do open heart surgery on himself.

Rule #14: Most managers succeed on the strength and skill of their staff.

Return To Table Of Contents

Initial Work

Rule #15: The seeds of problems are laid down early. Initial planning is the most vital part of a project. The review of most failed projects or project problems indicate the disasters were well planned to happen from the start.

Return To Table Of Contents

Communications

Rule #16: Cooperative efforts require good communications and early warning systems. A project manager should try to keep his partners aware of what is going on and should be the one who tells them first of any rumor or actual changes in plan. The partners should be consulted before things are put in final form, even if they only have a small piece of the action. A project manager who blindsides his partners will be treated in kind and will be considered a person of no integrity.

Rule #17: Talk is not cheap; but the best way to understand a personnel or technical problem is to talk to the right people. Lack of talk at the right levels is deadly.

Rule #18: Most international meetings are held in English. This is a foreign language to most participants such as Americans, Germans, Italians, etc. It is important to have adequate discussions so that there are no misinterpretations of what is said.

Rule #19: You cannot be ignorant of the language of the area you manage or with that of areas with which you interface. Education is a must for the modern manager. There are simple courses available to learn computerese, communicationese and all the rest of the modern "ese's" of the world. You can't manage if you don't understand what is being said or written.

Return To Table Of Contents

People

Rule #20: You cannot watch everything. What you can watch is the people. They have to know you will not accept a poor job.

Rule #21: We have developed a set of people whose self interest is more paramount than the work or at least it appears so to older managers. It appears to the older managers that the newer ones are more interested in form than in substance. The question is are old managers right or just old? Consider both viewpoints.

Rule #22: A good technician, quality inspector, and straw boss are more important in obtaining a good product than all the paper and reviews.

Rule #23: The source of most problems is people, but darned if they will admit it. Know the people working on your project to know what the real weak spots are.

Rule #24: One must pay close attention to workaholics—if they get going in the wrong direction, they can do a lot of damage in a short time. It is possible to overload them and cause premature burnout but hard to determine if the load is too much, since much of it is self generated. It is important to make sure such people take enough time off and that the workload does not exceed 1 1/4 to 1 1/2 times what is normal.

Rule #25: Always try to negotiate your internal support at the lowest level. What you want is the support of the person doing the work, and the closer you can get to him in negotiations the better.

Rule #26: If you have someone who doesn't look, ask, and analyze; ask them to transfer.

Rule #27: Personal time is very important. You must be careful as a manager that you realize the value of other people's time (i.e., the work you hand out and meetings should be necessary). You must, where possible, shield your staff from unnecessary work (i.e., some requests should be ignored or a refusal sent to the requestor).

Rule #28: People who monitor work and don't help get it done never seem to know exactly what is going on (being involved is the key to excellence).

Rule #29: There is no greater motivation than giving a good person his piece of the puzzle to control, but a pat on the back or an award helps.

Rule #30: It is mainly the incompetent that don't like to show off their work.

Rule #31: There are rare times when only one man can do the job. These are in technical areas that are more art and skill than normal. Cherish these people, but get their work done as soon as possible. Getting the work done by someone else takes two or three times longer and the product is normally below standard.

Rule #32: People have reasons for doing things the way they do them. Most people want to do a good job and, if they don't, the problem is they probably don't know how or exactly what is expected.

Rule #33: If you have a problem that requires additional people to solve, you should approach putting people on like a cook who has under-salted the food.

Return To Table Of Contents

Reviews and Reports

Rule #34: NASA has established a set of reviewers and a set of reviews. Once firmly established, the system will fight to stay alive, so make the most of it. Try to find a way for the reviews to work for you.

Rule #35: The number of reviews is increasing but the knowledge transfer remains the same; therefore, all your charts and presentation material should be constructed with this fact in mind. This means you should be able to construct a set of slides that only needs to be shuffled from presentation to presentation.

Rule #36: Hide nothing from the reviewers. Their reputation and yours is on the line. Expose all the warts and pimples. Don't offer excuses—just state facts.

Rule #37: External reviews are scheduled at the worst possible time, therefore, keep an up-to-date set of business and technical data so that you can·rapidly respond. Not having up-to-date data should be cause for dismissal.

Rule #38: Never undercut your staff in public (i.e., In public meetings, don't reverse decisions on work that you have given them to do). Even if you direct a change, never take the responsibility for implementing away from your staff.

Rule #39: Reviews are for the reviewed an not the reviewer. The review is a failure if the reviewed learn nothing from it.

Rule #40: A working meeting has about six people attending. Meetings larger than this are for information transfer (management science has shown that, in a group greater than twelve, some are wasting their time).

Rule #41: The amount of reviews and reports are proportional to management's understanding (i.e., the less management knows or understands the activities, the more they require reviews and reports). It is necessary in this type of environment to make sure that data is presented so that the average person, slightly familiar with activities, can understand it. Keeping the data simple and clear never insults anyone's intelligence.

Rule #42: Managers who rely only on the paperwork to do the reporting of activities are known failures.

Rule #43: Documentation does not take the place of knowledge. There is a great difference in what is supposed to be, what is thought to have happened, and reality. Documents are normally a static picture in time that get outdated rapidly.

Rule #44: Just because you give monthly reports, don't think that you can abbreviate anything in a yearly report. If management understood the monthlies, they wouldn't need a yearly.

Rule #45: Abbreviations are getting to be a pain. Each project now has a few thousand. This calls on senior management to know hundreds. Use them sparingly in presentations unless your objective is to confuse.

Rule #46: Remember, it is often easier to do foolish paperwork that to fight the need for it. Fight only if it is a global issue which will save much future work.

Return To Table Of Contents

Contractors and Contracting

Rule #47: A project manager is not the monitor of the contractor's work but is to be the driver. In award fee situations, the government personnel should be making every effort possible to make sure the contractor gets a high score (i.e., be on schedule and produce good work). Contractors don't fail, NASA does and that is why one must be proactive in support. This is also why a low score damages the government project manager as much as the contractor's manager because it means that he is not getting the job done.

Rule #48: Award fee is a good tool that puts discipline both on the contractor and the government. The score given represents the status of the project as well as the management skills of both parties. The project management measurement system (pms) should be used to verify the scores. Consistent poor scores require senior management intervention to determine the reason. Consistent good scores which are consistent with pms reflect a well-run project, but if these scores are not consistent with the pms, senior management must take action to find out why.

Rule #49: Morale of the contractor's personnel is important to a government manager. Just as you don't want to buy a car built by disgruntled employees, you don't want to buy flight hardware developed by under-motivated people. You should take an active role in motivating all personnel on the project.

Rule #50: Being friendly with a contractor is fine—being a friend of a contractor is dangerous to your objectivity.

Rule #51: Remember, your contractor has a tendency to have a one-on-one interface with your staff. Every member of your staff costs you at least one person on the contract per year.

Rule #52: Contractors tend to size up the government counterparts and staff their part of the project accordingly. If they think yours are clunkers, they will take their poorer people to put on your project.

Rule #53: Contractors respond well to the customer that pays attention to what they are doing but not too well to the customer that continually second-guesses their activity. The basic rule is a customer is always right but the cost will escalate if a customer always has things done his way instead of how the contractor planned on doing it. The ground rule is: never change a contractor's plans unless they are flawed or too costly (i.e., the old saying that better is the enemy of good).

Rule #54: There'is only one solution to a weak project manager in industry—get rid of him fast. The main job of a project manager in industry is to keep the customer happy. Make sure the one working with you knows that it is not flattery but on-schedule, on-cost, and a good product that makes you happy.

Return To Table Of Contents

Engineers and Scientists

Rule #55: Over-engineering is common. Engineers like puzzles and mazes. Try to make them keep their designs simple.

Rule #56: The first sign of trouble comes from the schedule or the cost curve. Engineers are the last to know they are in trouble. Engineers are born optimists.

Rule #57: The project has many resources within itself. There probably are five or ten system engineers considering all the contractors and instrument developers. This is a powerful resource that can be used to attack problems.

Rule #58: Many managers, just because they have the scientists under contract on their project, forget that the scientists are their customers and many times have easier access to top management than the managers do.

Rule #59: Most scientists are rational unless you endanger their chance to do their experiment. They will work with you if they believe you are telling them the truth. This includes reducing their own plans.

Return To Table Of Contents

Hardware

Rule #60: In the space business, there is no such thing as previously flown hardware. The people who build the next unit probably never saw the previous unit. There are probably minor changes (perhaps even major changes); the operational environment has probably changed; the people who check the unit out in most cases will not understand the unit or the test equipment.

Rule #61: Most equipment works as built, not as the designer planned. This is due to layout of the design, poor understanding on the designer's part, or poor understanding of component specifications.

Return To Table Of Contents

Computers and Software

Rule #62: Not using modern techniques, like computer systems, is a great mistake, but forgetting that the computer simulates thinking is a still greater mistake.

Rule #63: Software has now taken on all the parameters of hardware (i.e., requirement creep, high percentage of flight mission cost, need for quality control, need for validation procedures, etc.). It has the added feature that it is hard as blazes to determine it is not flawed. Get the basic system working first and then add the bells and whistles. Never throw away a version that works even if you have all the confidence in the world that the newer version works. It is necessary to have contingency plans for software.

Rule #64: Knowledge is often revised by simulations or testing, but computer models have hidden flaws not the least of which is poor input data.

Rule #65: In olden times, engineers had hands-on experience, technicians understood how the electronics worked and what it was supposed to do, and layout technicians knew too—but today only the computer knows for sure and it's not talking.

Return To Table Of Contents

Senior Management, Program Offices, and Above

Rule #66: Don't assume you know why senior management has done something. If you feel you need to know, ask. You get some amazing answers that will astonish you.

Rule #67: Know your management—some like a good joke, others only like a joke if they tell it.

Rule #68: Remember the boss has the right to make decisions. Even if you think they are wrong, tell the boss what you think but if he still wants it done his way; do it his way and do your best to make sure the outcome is successful.

Rule #69: Never ask management to make a decision that you can make. Assume you have the authority to make decisions unless you know there is a document that states unequivocally that you can't.

Rule #70: You and the Program Manager should work as a team. The Program Manager is your advocate at NASA HQ and must be tied into the decision makers and should aid your efforts to be tied in also.

Rule #71: Know who the decision makers on the program are. It may be someone outside who has the ear of Congress or the Administrator, or the Associate Administrator, or one of the scientists—someone in the chain of command—whoever they are. Try to get a line of communication to them on a formal or informal basis.

Return To Table Of Contents

Program Planning, Budgeting, and Estimating

Rule #72: Today one must push the state of the art, be within budget, take risks, not fail, and be on time.

Strangely, all these are consistent as long as the ground rules such as funding profile and schedule are established up front and maintained.

Rule #73: Most of yesteryear's projects overran because of poor estimates and not because of mistakes. Getting better estimates will not lower costs but will improve NASA's business reputation. Actually, there is a high probability that getting better estimates will increase costs and assure a higher profit to industry unless the fee is reduced to reflect lower risk on the part of industry. A better reputation is necessary in the present environment.

Rule #74: All problems are solvable in time, so make sure you have enough schedule contingency—if you don't, the next project manager that takes your place will.

Rule #75: The old NASA pushed the limits of technology and science; therefore, it did not worry about requirements creep or overruns. The new NASA has to work as if all projects are fixed price; therefore, requirement creep has become a deadly sin.

Rule #76: Know the resources of your center and, if possible, other centers. Other centers, if they have the resources , are normally happy to help. It is always surprising how much good help one can get by just asking.

Rule #77: Other than budget information prior to the President's submittal to Congress, there is probably no secret information on a project—so don't treat anything like it is secret. Everyone does better if they can see the whole picture so don't hide any of it from anyone.

Rule #78: NASA programs compete for budget funds—they do not compete with each other (i.e., you never attack any other program or NASA work with the idea that you should get their funding). Sell what you have on its own merit.

Rule #79: Next year is always the year with adequate funding and schedule. Next year arrives on the 50th year of your career.

Return To Table Of Contents

The Customer

Rule #80: Remember who the customer is and what his objectives are (i.e., check with him when you go to change anything of significance).

Return To Table Of Contents

NASA Management Instructions

Rule #81: NASA Management Instructions were written by another NASA employee like you; therefore, challenge them if they don't make sense. It is possible another NASA employee will rewrite them or waive them for you.

Return To Table Of Contents

Decisionmaking

Rule #82: Wrong decisions made early can be recovered from. Right decisions made late cannot correct them.

Rule #83: Sometimes the best thing to do is nothing. It is also occasionally the best help you can give. Just listening is all that is needed on many occasions. You may be the boss, but if you constantly have to solve someone's problems, you are working for him.

Rule #84: Never make a decision from a cartoon. Look at the actual hardware or what real information is available such as layouts. Too much time is wasted by people trying to cure a cartoon whose function is to explain the principle.

Return To Table Of Contents

Professional Ethics and Integrity

Rule #85: Integrity means your subordinates trust you.

Rule #86: In the rush to get things done, it's always important to remember who you work for. Blindsiding the boss will not be to your benefit in the long run.

Return To Table Of Contents

Project Management and Teamwork

Rule #87: Projects require teamwork to succeed. Remember, most teams have a coach and not a boss, but the coach still has to call some of the plays.

Rule #88: Never assume someone knows something or has done something unless you have asked them; even the obvious is overlooked or ignored on occasion, especially in a high stress activity.

Rule #89: Whoever said beggars can't be choosers doesn't understand project management, although many times it is better to trust to luck than to get poor support.

Rule #90: A puzzle is hard to discern from just one piece; so don't be surprised if team members deprived of information reach the wrong conclusion.

Rule #91: Remember, the President, Congress, OMB, NASA HQ, senior center management, and your customers all have jobs to do. All you have to do is keep them all happy.

Return To Table Of Contents

Treating and Avoiding Failures

Rule #92: In case of a failure:

- a) Make a timeline of events and include everything that is known.
- b) Put down known facts. Check every theory against them.
- c) Don't beat the data until it confesses (i.e., know when to stop trying to force-fit a scenario).
- d) Do not arrive at a conclusion too fast. Make sure any deviation from normal is explained. Remember the wrong conclusion is prologue to the next failure.
- e) Know when to stop.

Rule #93: Things that fail are lessons learned for the future. Occasionally things go right: these are also lessons learned. Try to duplicate that which works.

Rule #94: Mistakes are all right but failure is not. Failure is just a mistake you can't recover from; therefore, try to create contingency plans and alternate approaches for the items or plans that have high risk.

Rule #95: History is prologue. There has not been a project yet that has not had a parts problem despite all the qualification and testing done on parts. Time and being prepared to react are the only safeguards.

Rule #96: Experience may be fine but testing is better. Knowing something will work never takes the place of proving that it will.

Rule #97: Don't be afraid to fail or you will not succeed, but always work at your skill to recover. Part of that skill is knowing who can help.

Rule #98: One of the advantages of NASA in the early days was the fact that everyone knew that the facts we were absolutely sure of could be wrong.

Rule #99: Redundancy in hardware can be a fiction. We are adept at building things to be identical so that if one fails, the other will also fail. Make sure all hardware is treated in a build as if it were one of a kind and needed for mission success.

Rule #100: Never make excuses; instead, present plans of actions to be taken.

Return To Table Of Contents

Valuating Human Capital

Introduction

Knowledge management has been gaining worldwide attention in recent years. Many organizations have already created a new position of "Chief Knowledge Officer" to help better manage, share, create, secure, and distribute their knowledge-based assets. As this field matures, it is critical to develop some measures (and methodologies) for valuating knowledge assets. This section will highlight some of the leading work in this area, describe factors affecting human capital growth and their relationships, and discuss some future directions necessary to improve the current state of the art.

Measures for Valuating Knowledge Assets

There are several leading organizations and consultants that have been developing measures and techniques for valuating knowledge assets. Some of this work is highlighted below:[1-8]

Skandia

- Financial Focus: Gross premium income; insurance result
- Customer Focus: Satisfied customer index; customer loyalty; market share
- Human Focus: Number of employees; average age; empowerment index
- Process Focus: Operating expense ratio; premium income/salesperson; net claims ratio
- Renewal and Development: Training expense/employee; sales-oriented operations

Dow Chemical

- Projected costs until expiration
- Percentage of annual intellectual asset management (IAM) costs of R&D budget
- Ratio of NPV apportioned to intellectual assets (IA) to net present cost of R&D per period
- Percentage of competitive samples analyzed that initiate business actions by purpose
- Percentage of "Business Using"
- Percentage of "Business Will Use" more than 5 years since priority filing
- Quantitative Value Classification as a percentage of projected costs (e.g., what percentage of portfolio costs are for defensive cases, potential license cases, key cases)
- Classifications completed

Buckman Laboratories

- Percentage of company effectively engaged with customer (target: 80%)
- Percentage of revenues invested in knowledge transfer system
- Number of college graduates
- Sales of new products less than 5 years old as a percentage of total sales

Karl-Erik Sveiby (author of The New Organizational Wealth: Managing and Measuring Knowledge Based Assets)

- Developed the Intangible Assets Monitor to focus on external structure, internal structure, and competence of people

Ed Mahler (Dupont/E.G. Mahler and Company)

- Net training per employee plus R&D

Canadian Imperial Bank of Commerce

- Human capital (the skills individuals need to meet the customer needs)
- Structural capital (information required to understand specific markets)
- Customer capital (the essential data about the bank's customer base)

George Harmon (President of Micord Corporation)

$$Iv = (At-An)-(Lt-Ln)-(Ig+If+Ir+Id+It+Is+Iu)$$

where Iv = value of the particular information
At = the assets derived from the information at time of arrival
An = the assets if the information did not arrive
Lt = the liabilities derived from the information at time of arrival
Ln = the liabilities if the information did not arrive
Ig = the cost to generate the information
If = the cost to format the information
Ir = the cost to reformat the information
Id = the cost to duplicate the information
It = the cost to transmit or transport the information (distribute)
Is = the cost to store the information
Iu = the cost to use the information, including retrieval

Montague Institute (compilation)

- Relative value: progress, not a quantitative target, is the ultimate goal
- Balanced scorecard: supplements traditional financial measures with customers, internal business processes, and learning/growth
- Competency models: by observing and classifying the behaviors of "successful" employees and calculating the market value of their output, it is possible to assign a dollar value to the intellectual capital they create and use in their work
- Subsystem performance
- Benchmarking
- Business worth: evaluation focuses on the cost of missing or underutilizing a business opportunity, avoiding or minimizing a threat

- Business process auditing: measures how information enhances value in a given business process
- Knowledge bank: treats capital spending as an expense (instead of an asset) and treats a portion of salaries as an asset
- Brand equity valuation: measures the economic impact of a brand (or other intangible asset) on such things as pricing power, distribution reach, ability to launch new products, etc.
- Calculated intangible value: compares a company's return on assets with a published average return on assets for the industry
- Microlending: substitutes intangible "collateral" (peer group support, training, and the personal qualities of entrepreneurs) for tangible assets
- Colorized reporting: suggested by the Securities and Exchange Commission to supplement traditional financial statements with additional information (e.g., brand values, customer satisfaction measures, value of a trained work force, etc.)

Nuala Beck (Canadian economist)

- Knowledge ratio: expresses the number of knowledge workers as a percentage of total employment in an industry, individual company, or organization (measures the "Corporate IQ")
- Return-on-knowledge assets: the number of knowledge workers to profit earned
- Patent-to-stock price ratio: the ratio of the number of patents divided by the price of a company's stock
- Research-to-development ratio: the ratio of research dollars spent to the development dollars spent
- Research and development-to-patent ratio: the ratio of R&D investment to number of new patents issued

Factors Affecting Growth of Human Capital

If we define "knowledge assets" to be associated with only "human capital," we need to first develop a list of factors that affect human capital growth before we can develop a valuation methodology for measuring human capital. The following factors contribute to human capital growth:

- Formal training of employees
- R&D expenditures of the organization
- Morale (benefits, compensation, conferences, travel, vacation time, etc.)
- Formal education (i.e., degrees) of employees
- Mentoring and on-the-job training of employees
- Research skills
- Creativity and ingenuity
- Entre- and intra- preneurship skills
- Industry competition
- Half-life of information in industry
- Demand and supply of those in the field
- Retention rates of employees (studies indicate that retention is 20% of what we hear, 40% of what we see and hear, and 75% of what we see, hear, and do)
- Formalized knowledge transfer systems (e.g., lessons learned databases or best practices guidelines) institutionalized within the organization
- Informal knowledge transfer systems (e.g., speaking often with top management, secretaries and assistants to top management, attending company events, the "grapevine")
- Interaction with customers and users
- Stimulation and motivation (e.g., challenging assignments, giving responsibility and authority to the employee [employee empowerment], etc.)
- Physical environment and ambiance (e.g., nice office, reasonable resources, etc.)
- Internal environment within the organization (e.g., reasonableness of demands by management placed on the employees, etc.)
- Short-term (2–4 years) and long-term (5 years or more) prospects, from the employee's perspective, of the organization's viability and growth.

Developing Relationships Among These Human Capital Factors

Generally speaking, if these factors increase in a positive way, then human capital should expand (albeit with the possible exception of "industry competition" increasing). The converse is also true in that if these factors (discounting industry competition) take a negative turn, then human capital growth is likely to diminish or be stymied.

We can best group these 19 factors into the following categories:

- Training and Education (T&E)
  - Formal training of employees
  - Formal education of employees
  - Mentoring and on-the-job training
- Skills (S)
  - Research skills
  - Entre- and intra- preneurship skills
  - Retention rates
- Outside Pressures and Environmental Impacts (OP&EI)
  - Industry competition
  - Half-life of information in industry
  - Demand and supply of those in the field
- Internal and Organizational Culture (I&OC)
  - R&D expenditures of the organization
  - Formalized knowledge transfer systems
  - Informal knowledge transfer systems
  - Interaction with customers and users
  - Physical environment and ambiance
  - Internal environment within the organization
  - Short-term and long-term goals
- Psychological Impacts (PI)
  - Morale
  - Creativity and ingenuity
  - Stimulation and motivation

Some possible relationships and correlations could be posited between these five groups of factors affecting human capital growth (HCG):

$$HCG = T\&E + S + OP\&EI + I\&OC + PI$$

If T&E goes up, then S should generally increase. If T&E goes down, then S should generally decrease.

If I&OC is in a stressful state, then PI should go down. If I&OC is in a favorable state, then PI should increase.

If OP&EI increases, then it may be necessary to increase T&E and S expenditures. Also, if OP&EI increases, then it may create pressures on I&OC and thus possibly stifle PI. Alternatively, one may argue that if OP&EI

increases, then it may force employees and management to be more creative (to handle the competition) and thereby increase PI.

The Next Steps

In order to adequately determine human capital growth in an organization, empirical and exploratory testing needs to be conducted on these factors to test the completeness of this list of human capital growth factors and determine the possible relationships between and among these factors. A study of these measures needs to be conducted in order to determine how best to valuate and manage human capital and knowledge assets. From this research, some appropriate metrics for measuring human capital growth will be further developed and a methodology for valuating these knowledge assets will be created.

References

1. Bontis, N. (1996), "There's a Price on Your Head: Managing Intellectual Capital Strategically," *Business Quarterly,* Vol. 60, No. 4.
2. Brooking, A. (1996), *Intellectual Capital,* International Business, Thomson Press, London.
3. Bukowitz, W. and G. Petrash (1997), "Visualizing, Measuring, and Managing Knowledge," *Research-Technology-Management Journal,* Industrial Research Institute, Washington, D.C., July-August.
4. Edvinsson, L. and M. Malone (1997), *Intellectual Capital,* Harper Collins, New York.
5. Liebowitz, J. and L. Wilcox (Eds.) (1997), *Knowledge Management and Its Integrative Elements,* CRC Press, Boca Raton, FL.
6. Liebowitz, J. and T. Beckman (1998), *Knowledge Organizations: What Every Manager Should Know,* St. Lucie/CRC Press, Boca Raton, FL.
7. Montague Institute (1997), "Measuring Intellectual Capital," *Limited Edition* — newsletter on the Web, http://www.montague.com.
8. Sveiby, K. (1997), *The New Organizational Wealth: Managing and Measuring Knowledge-Based Assets,* Berrett-Koehler, San Francisco.

2 The Shoemaker's Children Go Barefoot

To Pedal Faster, Keep Your Shoes On!

Introduction

It is often the rule instead of the exception that organizations do not "practice what they preach." Whether it is the computer consulting firm that is not automated or the high tech firm that is not using the latest technology or the business school that can't manage its resources well, cases such as these are constantly surfacing. This is even true at the sole proprietorship level, as exemplified by the doctor who doesn't even get a physical himself each year.

It is often the case that organizations and individuals are not performing internally what they are practicing externally. This chapter will examine why this is so and will present some case studies that are evidences of this occurrence.

Factors That Contribute to the Shoemaker's Children Going Barefoot

There are several reasons why organizations and individuals may not practice what they preach. The first reason deals with time. Many organizations and individuals are so busy trying to make a profit and creating new business on the outside that they don't have time to improve their own operations internally. The doctor with the long office hours and who is often on call may not have the time to take proper care of himself. He is too busy healing others, instead of also protecting his own health by taking yearly physicals, exercising, and watching his diet. The telecommunications firm that is selling the latest and hottest services and products in telecommunications, but does not have the time to invest in an artificial intelligence/expert systems capability within the firm for diagnosing faults in networks, performing scheduling functions, and helping in other areas is another example of this phenomenon. Time seems to be a critical element in having the shoemaker's children go barefoot.

Another important factor contributing to this phenomenon is money. Some businesses may feel that they do not have the money to invest in improving their business operations. Some companies may not want to start up an expert systems group because they feel that the start-up costs for such a group are too high. This may not necessarily be a true statement, because with the development of PC-based expert system shells, an expert system may be developed for less than $25,000 instead of costing ten times that amount. For example, Dupont's average cost of their expert systems is $25,000, and the average return on investment of their expert systems is $100,000. So, it may be the case that the organization is not well-informed as to the possibilities of bringing in a new technology or capability to the firm. The company may think that the cost of developing such a capability is prohibitive, where in fact the cost may be much less than what is perceived. Of course, there are many "real" cases where the lack of financial resources or an expected poor payback on the investment may warrant a decision not to invest in improving a business operation within the firm.

The third factor for the shoemaker's children going barefoot is management's poor business acumen. Some managers may not have the foresight to invest in a new technology for improving the business operations of the firm. The dermatologist running her own practice may feel that it is not justifiable to have a computer in her office for word processing capabilities, cash management, appointment scheduling, and other functions, whereas, in reality, her business operations would be greatly facilitated by a computer in her office, at the very least for sending out patient referral letters, updating patient records, and keeping track of accounts receivable. In other cases, some large businesses may fall into the pitfall of being "in the railroad industry, not the transportation industry." The failure on the part of management to understand their industry and business scope may be one reason why their business efficiency and effectiveness are not improved.

The last major factor for companies not practicing what they preach is the coping with change phenomenon. Companies may feel uneasy with change; the thinking is, why try to improve our business operations if our operations are running smoothly as is. In other words, "if it's not broke, why fix it?" This may be a valid statement, but in the long run, it may be advantageous to the company to invest in a new technology to improve the way it does its business. There may be a way to enhance operations in the long run by changing some business practice or bringing on-line a new technology. Companies must properly evaluate their needs, and must determine how best to meet their strategic goals.

The next section illustrates some sanitized versions of situations where the shoemaker's children went barefoot.

Case Examples of Organizations Not Practicing What They Preach

This section will discuss three examples of organizations that have fallen into the trap of not practicing what they preach. The names of the organizations have been omitted for confidentiality purposes.

A Computer Manufacturing Firm

This first case involves a major international manufacturer of computer systems. This manufacturer developed and sold state-of-the-art hardware and software. It was selling computers with parallel processing capabilities and advanced software engineering techniques. The paradox of this situation was that even though the company was selling and promoting high technology products, their mode of business operations was somewhat antiquated.

Several examples within the company are evidences of this point. First, even though Lisp software was being sold for several years to customers for developing artificial intelligence/expert system applications, the company itself did not have an AI/expert system group, nor was it using expert system technology in its business operations. The company was manufacturing computers that could be used for AI applications, but the company was not exploring how expert systems could help its own operations. This is a good example of the shoemaker's children going barefoot.

Another evidence of this phenomenon within the computer manufacturing company is that it was building computers that could be used for customer computer-integrated manufacturing (CIM) operations, but the computer manufacturer itself did not use CIM for building these computers. This paradox was also perpetuated in the lack of integrated systems within the company. Even though the company advocated that its customers should have integrated systems, the company itself was not a role model in this regards. Again, the cliche of not practicing what one preaches is evident here.

A Military Laboratory

Another case example of the shoemaker's children going barefoot involves a military research laboratory. This particular military laboratory was developing sophisticated techniques, equipment, and applications in physics, chemistry, computer science, space science, and other areas. The shoemaker's children paradox struck the laboratory in the lack of a management information system for handling the generation and tracking of procurement actions. Even though the laboratory was developing sophisticated software for military applications, it did not internally develop or use software for improving its business operations for specifically easing the development of procurement request packages. Software was not being used in the laboratory that would generate the various procurement forms based on a fill-in-the-blank procedure by the user. As a result of not having this capability, the generation of procurement request packages was difficult to accomplish, because there was no easy way of comparing the forms for a new proposed procurement action to the forms used in similar procurement actions in the past. Again, the shoemaker's children were going barefoot because not enough attention was being devoted to improving the internal business operations of the organization.

The Systems Consulting Firm

A major international management and systems consulting firm ran into the same shoemaker dilemma. This company performed management studies and developed information systems for large organizations. Even though the company would recommend how other companies should run, the company itself did not take a close enough look at its own internal procedures. Specifically, there was an artificial intelligence group that existed in two divisions of the company. Even though these two groups were separated by only one floor between them, there was little or no communication or pooling of resources being accomplished between these two groups. Thus, one group might have embarked on an expert systems project that might have been similar to one done in the other group, but because of the lack of communication between these groups, the wheel might have been reinvented. The major problem that existed was that each group was a separate profit-seeking center, and there was competition between these two similar groups within the same company. Here again, the systems consulting firm would recommend to other companies how to manage their resources, but the consulting firm seemed to be doing a poor job in managing its own resources. Thus, the shoemaker's children dilemma strikes again.

Remedies for Avoiding the Shoemaker's Paradox

There are several ways that companies can plan in advance to avoid the possible shoemaker's paradox. The first way is to have a steering committee within the company whose major responsibility is to look at the processes and business practices within the firm and determine how these processes and practices should be improved. This committee should meet regularly and should be made up of the strategically focused managers representing the various business constituencies at the company. This committee would also provide guidance on planning and priority setting for the company in the context of a multidisciplinary view looking out several years. There should also be an internal audit team to later check if the implemented business practices recommended by this committee are actually being done. A second way to potentially avoid the shoemaker's dilemma is to have the company automatically invest annually a certain amount of dollars towards improving the business operations within the firm. The company might want to have a contest and offer a monetary prize to the employee who suggests the best way of improving the business operations within the company (a cost/benefit analysis should also be conducted by the employee). Another way of avoiding the shoemaker's dilemma is to keep the management and staff of the company up to date on new technologies and management theory. A seminar series could be conducted on a monthly basis where speakers from within and outside the company would give presentations on the latest technologies and management practices that could impact the firm. The last way for possibly avoiding the shoemaker's paradox is to encourage discussions among various groups within the company. If possible, try to set up cross-departmental consulting teams in order to facilitate some technology transfer and exchange of information/ideas among company individuals and departments. By keeping company individuals informed and allowing for interaction among different departments, new ideas might be generated on how to improve the business operations of the firm. A knowledge repository of best practices and lessons learned over the company's intranet would also help to avoid "going barefoot."

Creating an Awareness

The key to success in the information technology field is making sure that you "solve the business problem." This implies that you shouldn't force-fit technologies to problem areas. Rather, you should understand the scope and requirements of the problem or opportunity and then use the most appropriate technologies for best solving the problem at hand. All too often, information systems developers become so enamored with a particular technology that they begin to engage in the "hammer and the nail" phenomenon. Namely, if they only know how to use a hammer, every problem looks like a nail! This is one of the major causes of why some information technology projects fail.

In order to solve the problem and use the appropriate information systems technologies, the first step that is needed involves "creating an awareness of what the technologies can offer in solving the problem." There are a number of ways to create this awareness. These will be briefly discussed next.

Competition

One of the most fruitful ways of opening the eyes of management is to show how the competitors are successfully solving problems similar to those in the organization. In order to keep pace or stay ahead of its competition, the organization must realize how its competitors are maximizing their use of technology to solve related problems.

In one situation (that is described in this book), a computer manufacturer had a severe problem with configuring its customized computer systems for its clients. It was shown to upper management that their competitors at IBM, Hitachi, Siemens, Philips, Fujitsu, DEC, Motorola, and others had already successfully solved this problem through using expert systems technology. In fact, the enormous payback from doing this at DEC was published in the open literature. This helped management to gauge the value of using this technology in solving the problem, although in this case there were too many organizational barriers that prevented this technology and project from starting.

Some companies, like IBM with respect to the PC, prefer to take a wait-and-see approach to determining the marketplace, avoiding being the one taking most of the risks as the first one to enter the market. With the personal computer, IBM entered the market after their competitors had already tested the market and IBM soon became the world leader with the largest market share in this PC market.

Bottom Line

Another important way to create an awareness is to show management how solving the problem through recommended information technologies can result significantly in "the bottom line." Preparing a cost/benefit analysis and associated risk assessment can greatly sway management in determining whether the project should be accepted. Both tangible and intangible benefits need to be quantified as best they can. If the project doesn't hit home at increasing the organization's profits or decreasing its costs, then management may be unconvinced of the value of the project and put it on the back burner.

In the case of the configuration management problem mentioned above, management failed to see the "big picture" in having an expert configuration system dramatically decrease manufacturing costs and improve customer loyalty and satisfaction. Management felt there were higher priorities, which might have been true at the time.

Top-Down and Bottom-Up Approach

To create an awareness of the project and associated technologies, a top-down and bottom-up approach may be used. Top-down refers to gaining the enthusiasm of the CEO, then the Senior VP, other vice presidents, senior management, middle management, and then the users. Bottom-up refers to getting the users excited about how their problem will be solved. They in turn tell their managers, who in turn tell top management, and the enthusiasm and awareness bubbles up to the top through the organizational levels. Combining these two approaches allows a two-way process to develop whereby a ripple effect permeates the organization, thereby resulting in a greater awareness of the project and hopefully gaining project acceptance.

Real Problem That Needs Fixing

The development team (particularly the project leader and "project champion" in management) must convince senior management that the problem (and solution) being addressed is a real problem that needs fixing. The problem must be significant, tractable, and solvable. The computer configuration problem that shows over 90% of the configurations were done incorrectly the first time and it typically took 12 times to get the correct computer configuration is alarming. The company will have difficulty staying in business with these kinds of numbers. Convincing management that this is a "real" problem and proposing a reasonable way of solving this problem are important steps in the awareness creation process and in ultimate project acceptance.

Subtle Reminders

The use of subtle reminders may be another way to trigger awareness of a problem for needed project action. You can use some creative ways of achieving subtleness. For example, in the computer configuration case, the organization sent pads with the embossed motto "Artificial Intelligence is for Real" to senior management in the company. 3M also used this approach via printing a saying on their Post-It notes and circulating them to top management. When the manager would use a Post-It note, he/she would see what was printed on it and this would serve as a subtle (and typically, frequent) cue. As Barbara Bouldin points out in her book, *Agents of Change: Managing the Introduction of Automated Tools,* every management level that is committed results in a higher probability of obtaining the necessary resources, as well as generally securing the future of the change effort.

Summary

Creating an awareness of the problem and proposing a workable solution are important first steps for project inception. Gaining management and user commitment are critical components of your information technology project. Consider using some of the techniques described above to pave the way for project acceptance and ultimate support.

3 CESA:
An Expert System for Aiding in U.S. Defense Research Contracting

Just when you think you know the contracting rules, they change them on you!

Introduction

The contracting area is a ripe application for expert system development. In the U.S. defense contracting business, there are two major phases of a contract: a pre-award phase and a post-award stage. The contract pre-award phase involves the packaging of necessary forms and information to generate an acceptable procurement request package. The procurement request package typically serves as the vehicle for generating proposals in response to the government's advertised need. Once proposals are received, reviewed, and evaluated by an evaluation team within the sponsoring government activity, the successful bidder (i.e., contractor) is informed of his selection and some negotiations may follow to iron out any uncovered details. Once the contract is in place, it is monitored by a government representative. The monitoring of the contract and the reviewing of the work performed under the contract are part of the contract post-award phase.

Expert systems could play valuable roles in assisting in the contracting process. For example, expert systems could be used to help train new contracts specialists, help the Acquisition Request Originator (ARO) assemble a "complete" procurement request package, and then assist in tracking the processing of that package. Expert systems could also aid the Contracting Officer Technical Representative (COTR) in monitoring a contract, and provide advice if a deliverable is late or inferior to what was promised. An expert system might also act as an advisory system to the COTR to explain such areas as how to exercise an option, how to terminate a contract, or how to select appropriate remedies short of termination. Additionally, an expert system could help flag discrepancies or inconsistencies in contractor monthly progress reports.

This chapter discusses the development of an expert system called CESA (COTR Expert System Aid), constructed at a major U.S. military scientific laboratory.

Need for CESA

One of the most important advantages of expert systems technology is its ability to aid in the training function of a particular application. Before an expert retires or leaves an organization, an expert system can serve in capturing the expert's knowledge and experiential learning to help train others in performing a specific task. In this manner, individuals in the organization, especially neophytes, can learn from the successes and failures of their predecessors in order to improve their effectiveness and productivity.

In most government contracting, there is a myriad of skills involved in the contracting process. Some of these skills range from the knowledge required in handling procurement request generation and execution to monitoring and evaluating contractor performance. The front end of the process is the pre-award phase of the contract, in which the appropriate materials are packaged to support the proposed acquisition. The back end of the cycle is the post-award phase, which involves the monitoring of contractor performance and inspection and acceptance of goods and services received. The individual who is responsible for providing all technical requirements, specifications, justifications, and statements necessary to support the proposed acquisition and who usually helps evaluate the technical aspects of all proposals, bids, or quotations is designated the ARO. Thus, the ARO is concerned with the contract pre-award phase. The person in direct support of the contract post-award phase is the COTR. The COTR's duties include assuring quality, providing technical direction as necessary with respect to the specifications or statement of work, monitoring the progress, cost, and quality of contractor performance, and certifying invoices. At NRL, the individual who is certified as the COTR on a particular contract most likely also served as the ARO.

There are two major reasons for developing an expert system to help the ARO/COTR. First, the nature of contracting involves many complex and often changing rules and regulations. It is difficult for the ARO/COTR to remember and to keep up-to-date with these new rules and procedures, particularly since he/she is principally a scientist or engineer and not a contract specialist, and may be called upon to perform contract-related duties on an irregular basis. Even though the COTR takes a formal course and passes a test to become eligible for certification, an expert system can act as supplemental training and also can refresh the COTR on specific aspects of this material as necessary later on. A second reason for developing an ARO/COTR expert system aid is to provide an interactive, interesting way for the ARO/COTR to reference and learn contract information needed, as well as to furnish a convenient vehicle for assisting in ARO/COTR problem solving. It should be noted that the need for a system such as CESA at the military scientific laboratory was first raised by a member of its COTR community.

The next section will discuss the development process of CESA.

Development of CESA

CESA's development followed the rapid prototyping approach using the following knowledge engineering steps: problem selection, knowledge acquisition, knowledge representation, knowledge encoding, and knowledge testing and evaluation. Each of these steps will be discussed in turn.

Problem Selection

In response to a suggestion by a COTR at the laboratory that expert systems technology might be applied to help the COTR better perform his functions, a feasibility study was conducted and identified four possible alternatives for system development within the COTR environment. These were

- An expert system prototype for procurement request generation and routing;
- An expert system prototype to act as a training aid;
- An expert system prototype for specific problem-solving activities relating to the performance of a contract; and
- An expert system prototype for monitoring the progress of a contract.

Analysis using the analytic hierarchy process revealed that the areas of (1) COTR problem-solving activities relating to contract performance and (2) procurement request generation and routing appeared particularly amenable to expert system development at the laboratory.

The top-ranked COTR problem-solving alternative had to be scoped much further to select a well-bounded task for developing the expert system. After discussions with numerous individuals, especially our contracts expert who had over 27 years of contracting experience, it was decided that the contract pre-award phase (i.e., procurement request generation) would be a better task area for developing the expert system than would the post-award phase. The main reason for selecting the pre-award area as our focus was that it seems to be particularly troublesome to the ARO community at the laboratory; experience shows that contracting specialists often receive incomplete or inaccurate procurement request packages that have to be returned to the ARO before processing, thus delaying the procurement process. Based on the strong need for such a system in the pre-award area and the structure and specificity of the pre-award domain, it was decided that the pre-award phase would be a high-interest, high-payoff area for near-term expert systems development. If this version of CESA was successful, tasks within the post-award area could later be incorporated into CESA.

The pre-award area also met the criteria for selecting an appropriate problem for expert system development. Namely, it was mostly symbolic in nature; there was a general consensus on the correct solution; the problem took a few minutes to a few hours to solve; it was performed frequently; it did not require much common sense reasoning; it dealt mainly with cognitive skills as opposed to motor skills; and an expert existed who would be able to participate in the project. For these reasons, the pre-award area, procurement request generation and routing, was selected as an appropriate task for CESA. The next step in CESA's development involved knowledge acquisition. To prepare for this step, the CESA development team obtained various laboratory contracting instruction and handbooks to become more familiar with the pre-award domain before meeting with a contracts expert. By reviewing this documentation, the CESA development team felt more comfortable in conversing with the expert and asking the right questions.

Knowledge Acquisition

After selecting a well-bounded task, the next major step in developing CESA was performing the knowledge acquisition process. Knowledge was acquired through two major sources. The first source was the many laboratory instructions and handbooks that address the pre-award contract phase. The second major method of acquiring knowledge was through interviewing a contracts expert.

The expert was a retired annuitant who had over 27 years of contracting experience. Tasked by the head of the Contracts Division to take part in CESA's development, she was very enthusiastic about helping in this project because she felt there was a great need for developing such a system to assist ARO/COTRs at the laboratory. She also was excited that her expertise would be "preserved" and used to help others at the laboratory. The knowledge acquisition sessions were conducted with the expert once a week, 2 to 3 hours per meeting, in order to build CESA. The first session concentrated on identifying the major pre-award areas that are of concern to the contracts specialists and the ARO. After the first two meetings with the expert, it was determined that the major areas of concern in the pre-award phase could be decomposed as follows:

- Adequacy of the procurement request (PR) package
  - What is needed in a PR package
  - Justification and approval (J&A) if requirement to be specified is sole source
  - What type of contract is desired
  - The statement of work (SOW)
  - Evaluation procedures
  - Synopsis procedures

- Routing of either the advance copy of the procurement planning document (PPD) or PR package, or routing of the original PPD, or routing of the original PR
- Use of the PPD

In acquiring knowledge from the expert, various forms of interviewing methods were used. Structured interviews were effective because once the major pre-award areas were mapped out, the knowledge engineering team would acquire knowledge from the expert in each of these areas, one at a time. The technique of using "constrained information tasks" forced the expert to use her thinking within a short period of time, in order for the knowledge engineering team to determine the important heuristics involved. The technique of using "limited information" during parts of the interview required the expert to determine what was important, in terms of information being omitted and material that was used. Another useful interviewing technique used in acquiring knowledge for CESA was the scenario approach. In this method, the knowledge engineering team posed sample scenarios to the expert, and the expert would then "think aloud" during the process of solving these cases.

There were several factors that made the knowledge acquisition process an effective and enjoyable task. First, the expert allowed the knowledge engineering team to tape the knowledge engineering sessions. This was useful in obtaining a better, more complete record of discussions with the expert than simply jotting down some notes. Also, the expression and intonation of the expert, which had some significance in the phrasing of certain information, could be captured. Second, using the different interviewing techniques previously mentioned helped uncover a variety of knowledge that needed to be incorporated in CESA. Third, documentation such as the instructions and contracting handbooks was extremely useful in providing additional information for the knowledge base, as well as familiarizing the knowledge engineering team with the domain, as noted earlier, prior to interviewing the expert. Fourth, each knowledge acquisition session usually ran about 2 hours, which seemed to be an appropriate length of time for not tiring the expert. Fifth, as the knowledge engineering team climbed the learning curve in understanding the contracts area, the expert became more and more enthusiastic about the project. When the expert obtained hands-on experience early on in running versions of the CESA prototype, the expert really felt that CESA was "her baby."

After acquiring the knowledge for CESA, the next major step involved representing the acquired knowledge. This will be explained next.

Knowledge Representation

Knowledge in expert systems should be represented in the most natural way that the expert employs his knowledge. From the knowledge acquisition sessions with the expert, it became apparent that the expert's knowledge could be easily represented as condition–action rules, or IF–THEN rules. This notion was further confirmed by reviewing the documentation on contracting, which typically consisted as a series of IF–THEN clauses. For example, an excerpt from the COTR Handbook reads:

> "If the COTR finds that the contractor is not complying with a specific requirement contained in the contract, the COTR should call the contractor's attention to the discrepancy and seek voluntary commitment to remedy the failure. If the contractor makes such a commitment, the COTR should follow up to see if the action is taken."

The final version of CESA consisted of 243 rules. After representing the knowledge as rules, the next step was to encode the knowledge in the knowledge base. This step will be explained next.

Knowledge Encoding

To help in CESA's development process, Exsys Professional was selected as the expert system shell. Exsys was the shell selected for CESA's development because

- Exsys can handle rules, and it can use backward or forward chaining (backward chaining was needed for CESA's application);
- Exsys runs on the IBM PC, which was the recommended hardware for CESA since most ARO/COTRs have access to a PC;
- Exsys has a fairly easy-to-use text editor for creating the knowledge base; this feature was important because CESA eventually will need to be maintained by someone in Contracts who is not a computer specialist;
- Exsys can handle uncertainty in its rules;

- Exsys allows for free-text comments to provide definitions and descriptions of qualifiers and their values; and
- Exsys is a fairly inexpensive expert system shell.

The major drawback in using Exsys is its user interface; although adequate, it could certainly be improved in flexibility of displays. However, the documentation for Exsys is well written and, generally speaking, the advantages of using Exsys far outweighed the disadvantages.

Encoding the knowledge for CESA using Exsys was an iterative process. After acquiring and representing the knowledge for each pre-award area, the knowledge was subsequently encoded into the system, one pre-award area at a time. By quickly encoding prototypical cases into CESA, the expert could see some tangible results occurring from the knowledge acquisition sessions. Also, by encoding this knowledge early on, the expert could more easily identify omissions in the knowledge or incorrect knowledge being applied. By seeing how the chaining in CESA took place, the expert could easily identify if the proper conclusions were reached from the combinations of the input provided. When weaknesses in the knowledge base were identified, the knowledge was re-acquired, represented, and encoded into CESA.

To help develop and eventually maintain CESA, the knowledge base was set up in a modular fashion. Sections in the knowledge base related to each pre-award area, and within each pre-award area, further subdivisions are made. Thus, each pre-award area is fairly autonomous from the other pre-award areas; this means that there is little, if any, "interlinking" between pre-award areas, and knowledge revisions relating to one pre-award section can be made without affecting the knowledge in the other pre-award sections. Although this approach may add a seemingly redundant qualifier or two to some rules, it ultimately will help in the maintainance of CESA by Contracts personnel. The next step after encoding the knowledge was testing the knowledge and evaluating CESA.

Knowledge Testing and Evaluation

The last major step in developing CESA was knowledge testing and evaluation. Knowledge testing involved verification and validation of the knowledge. This relates to "developing the expert system right," and "developing the right expert system." Evaluation entails obtaining user feedback on the human factors design of the expert system, as well as commenting on the accuracy and quality of the decisions reached.

For CESA, verification and validation were performed in a variety of ways. First, CESA's knowledge was verified by checking all possible major paths for logical consistency. It turned out that some combinations of answers led to incorrect or incomplete conclusions. The knowledge base had to be augmented to cover these situations. Second, historical test cases were run against CESA to see if the expert system-generated results were the same as the results from the documented cases. Specifically, completed (and successfully processed) PR packages were used to see if CESA generated the same list of forms to be included in the PR package as were found in the successful PR packages. Generally speaking, CESA was very accurate in its judgments/conclusions. In fact, in one case, CESA indicated a form that should have been in the completed PR package but was not included. Third, the expert and other test users ran CESA with hypothetical cases to test the quality and accuracy of CESA's decisions. This greatly helped during the knowledge refinement process.

In terms of evaluating the "user-friendliness" of CESA, both naive users (those individuals who just completed the COTR course) and knowledgeable users (those persons who have been COTRs for a few years) were generally pleased with CESA, but suggested that additional free-text comments be added to better define some qualifiers and values used in the system. Also, some of the wording used in questions asked of the user needed to be improved, since some of the wording was contracts-specialist-oriented, as opposed to ARO/COTR-oriented. Additionally, some of the users would like to have seen a more graphically oriented user interface.

Epilogue

CESA was further tested and evaluated by numerous contracts experts and COTRs. To help in transitioning CESA to the Contracts Division, two contracts specialists (with computer experience) were selected by Contracts to maintain CESA once deployed. The CESA knowledge engineering team spent 8 weeks in classroom-style instruction (with labs and homework) in working with these two individuals. By the end of the 8 weeks, they felt comfortable in being able to maintain CESA.

CESA was then deployed into the field at the laboratory through distribution to the administrative officers at the laboratory. These individuals were responsible mostly for handling procurement and contracting activities within the various codes/departments in the laboratory. A briefing and overview of CESA were given to the administrative officers and any interested

COTRs, and then copies of CESA and user's guides were distributed to them. This activity was accomplished to coincide with the end of the government's fiscal year, where contracting activity in the fourth quarter is usually high in order to use up the money budgeted to the various codes. The hope was that this would be an opportune time for the COTRs to use CESA to help them in putting together their procurement request packages.

After several months, CESA was being used on a frequent basis. A new Commanding Officer was then selected to head up the laboratory, and a number of changes were being made in the contracting regulations. It appeared that CESA was not being maintained to reflect these new changes. Upon further investigation, the Head of Contracts (who was the champion of the CESA effort) had recently left the laboratory and the two contracts specialists assigned to maintaining CESA had really no incentives to continue to update it. This was apparent because a fundamental error was made: the two contracts specialists were not being evaluated on how well CESA was being maintained; in other words, as part of their annual job performance appraisal, there was no mention about maintaining CESA. Thus, since they weren't being "rewarded" for maintaining CESA as part of their busy jobs, then why do it? Eventually, CESA's accuracy had dropped due to the lack of maintenance of CESA and CESA faded into the sunset! An important lesson was learned in "empowerment and reward."

4 The Independent Office Supply Store Without Any Automation

Make Sure You Have All the Necessary Supplies and Tools!

Another example of the shoemaker's children going barefoot is the real story of an independent office supply store owner who sold computers, fax machines, and other automated equipment but never used automation for his own business operations. The owner of this store in recent years increasingly found it difficult to compete with the large, discounted office supply chains who had taken many customers away from this small office supply store with their low prices. The small business owner had to rely on service, even delivering one stapler to a customer 20 miles away. The small business owner had two employees besides himself. One employee typically worked the cash register, and the second employee handled the calls for orders and assembled them for later drop-off.

The owner of the store was the buyer, marketing representative, and even the delivery man. The walk-in customer traffic to the store was minimal, and the owner relied on servicing his customers with deliveries within the hour. Increased competition in this "servicing area" became evident as the large, discounted office supply stores also promised free prompt deliveries guaranteed (although, there was a minimum for the amount of goods purchased before a delivery would be made). Rent for the small office supply store was rising with the proposed new lease, and the small business owner started to contemplate whether he should just lease warehouse space at a cheaper rent, since his walk-in customer traffic had dwindled.

The owner never used any type of computerization to improve the efficiency (and ultimately the costs) of his business. He didn't use a computer to help inventory his supply. Periodically, he would walk around the store counting the number of each office item, instead of using the computer to track his inventory on a perpetual basis. He also rarely used his fax machine for his own business operations (such as faxing his suppliers for more office products), and he actually would keep his fax machine turned off unless a walk-in customer wanted to fax something. If he let the fax machine be turned on always, his "service customers" could fax orders to this store, but this apparently was not considered.

A computer could also help the owner keep track of his accounts payables and receivables. Also, a database could easily be set up in order to have customer information readily available for billing, deliveries, and mailings. Having a computer system should also help him become more literate about computers when customers ask him questions about computers in general.

So why didn't the owner invest in a computer system for his business? In speaking with the owner, he gave several reasons. First, he never really used a computer system and had fear and anxiety over the notion of using it, especially for something as important as helping him run his store operations. Second, since no one was competent to develop and use a computer system in his store, he would become dependent on a computer consultant or would have to hire someone else. The owner didn't feel comfortable with this notion. Third and probably most important, the owner didn't want to spend the money (especially in difficult times). The owner was rather tight with this money, even to the point of not running the air conditioning system in the store in hot Florida!

The end result of this story is that the owner has a very stubborn personality: he probably won't change anything until he goes out of business. The last thing that he would consider is a computer system for the store. He probably should concentrate on getting new clients, instead of spending the time acting as the delivery man. Also, he should strongly consider the idea of working out of a warehouse, since his customer walk-in traffic is so poor. In this manner, he would alleviate his large rent bills. Most likely and unfortunately, it is only a matter of time before the large, discounted office supply stores swallow him up.

5 A Tax Help Desk System in the Government

Did I Forget to Tell the Users How to Work It?

Introduction

Management is crucial. This point has been echoed by many, including a senior manager at Deloitte & Touche, who said, "Seldom have we seen system development projects fail because of the technology. Typically, failures can be attributed to management issues surrounding the problem or the project itself."

61

Along these lines, there are three kinds of people: those who "make" things happen, those who "watch" things happen, and those who "wonder" what happened.

The following short case study describes a project that was technically sound, but toward the implementation and technology transfer stages quickly became a nightmare. In fact, this system was eventually decommissioned.

The Scope of the Tax Help Desk System

In this organization under study, there are about 5,000 front-line assistors who help answer questions regarding completing tax forms. They do not provide tax advice, but help in responding to questions dealing with unclear terms associated with the government tax forms. These front-line assistors typically handle 40 million telephone inquiries, 18 million inquiries on tax law, 139 tax topic categories, 10–50 parameters per tax topic, and reference 159 official government tax publications. Indeed, this is an extremely burdensome and difficult task to handle this volume of information and calls.

In a government study of the performance of these front-line assistors, it turned out that 63% were correct and complete (gave the right answer for the right reason), 15% were correct but incomplete (gave the right answer for the wrong reason), and 22% were incorrect (gave the wrong answer).

In order to improve the performance of these front-line assistors, an expert system was developed to aid the assistors in fielding these tax-related calls. The objectives of this system were to

- Provide expertise to enable front-line assistors to directly answer technical tax questions beyond their expected scope of knowledge;
- Explain the system's reasoning during and after the consultation session;
- Justify the system's advice during and at the end of the session;
- Printout a hardcopy of the session to the taxpayer upon request;
- Serve as a training aid through use of training examples and through exploration of problem space.

The advantages of such an expert system would be to shift the burden of knowing, remembering, and reasoning about details of tax topics from the human mind to the expert system. Also, explanations would show how decisions and advice were determined and would reduce referral of simple questions to domain experts. Additionally, this system should reduce "answer shopping" by providing accurate, complete, and consistent advice. Other advantages include reducing requests for government tax publications by mailing print-outs of advice given and an explanation, and training assistors by exploring the system. A major disadvantage of using an expert system approach is the slow development of handcrafted knowledge bases and screen interfaces.

The typical knowledge engineering steps were used to develop this expert system, namely problem selection, knowledge acquisition, knowledge representation, knowledge encoding, knowledge testing and evaluation, and implementation and maintenance.

Problem selection involved task decomposition. This followed these steps:

- Understanding and clarifying questions;
- Classifying questions;
- Referring questions;
- Researching and referencing tax information;
- Recalling tax information;
- Matching tax law information to the taxpayer's situation;
- Answering questions;
- Explaining answers;
- Insuring that taxpayers understand answers.

Knowledge was acquired from domain experts and associated tax regulations, policies, and documentation, and the knowledge was represented in a rule-based format. In 4 years, the number of topics in the tax help desk expert system grew from 25 to about 400, and the number of users grew from 3 to about 400. The system was tested and had accuracy rates in the mid-90 percent range.

Then the expert system was transferred from the expert systems developers' laboratory to the user group in the organization. A section was established within the user group for this expert system. The laboratory provided 1 year of full support in testing, enhancing, and maintaining the expert system. They also trained user group personnel on how to use the expert system shell and how to maintain/update the system. After the first year, the laboratory would supply only consulting services if requested.

Epilogue

In the second year after the expert system was transferred to the user group, needed changes were made to the expert system's knowledge base without much user involvement. The changes were not well tested, and so the accuracy of the tax help desk expert system began to fall dramatically. Since the accuracy of the knowledge base dwindled, the field stopped using portions of the knowledge base. Eventually, as one might guess, the system fell into disuse and was gradually decommissioned.

Automation: Sink or Swim?

Lessons Learned

An information system, whether an expert system or otherwise, can be thought of as a "living, breathing thing." Once the system is developed, the life cycle does not stop there. Rather, maintenance of the system is crucial and actually results in much of the expense of the total system life cycle costs. Providing adequate training, some hand-holding, consulting services, and other mechanisms are necessary for helping to ensure a smooth and successful implementation and transition. In the case of the tax help desk system, perhaps the laboratory should have been working closely with the user organization even after 1 year of technology transfer. This would have ensured that the knowledge base would be updated on a regular and correct basis. Additionally, it would have been important to empower the individuals to maintain the system, and have incentives (e.g., merit pay, etc.) to promote its accuracy, maintenance, and use. A follow-up post-audit by the developing team perhaps a month after technology transfer would have helped to identify the need for providing a better way of performing maintenance on the system.

6 War Stories

This Should Work!!

Here are a few war stories about computer systems in small businesses and practices. These actual war stories are being told so that you can avoid the mistakes of others.

In The Trenches #1: Don't Always Trust a Relative

The first war story involves a small business that decided to use computers to help employees perform their parking management business activities. The company had about 12 people, of which the top two executives were family members.

When the company decided to get into the computerized age, they were lucky enough to have an in-law who had his own computer consulting firm set up the system for them. The in-law did this service free of charge, and he essentially did it on his own free time.

A couple of problems quickly set in after the company started using the computer system. First, there was little training on how to use the system. The users were not given documentation on the system or even documentation on the applications software that was put on the system. So, whenever there was a question or problem, the in-law would be called and he eventually would come in and fix the problem. Thus, a strong reliance on the in-law was firmly built. Second, there was no person within the company who was designated as the "systems analyst" or "computer manager" to oversee the computer. No one from within the company looked at getting new updates for their applications software, or looked at other software that might help out their company, or had some experience to correct system problems when they would occur. As a result, the company became even more dependent on the in-law. Third, the in-law did not put together the "best system," and as a result, the system would break down fairly often. This was particularly troublesome to the company because the firm strongly relied now on the computer system to help employees perform their daily duties.

Well, you can imagine what eventually ensued during the months after the computer system was initially installed. The company was pretty disenchanted with the system, as well as the in-law. One major problem with the in-law was that he would not come over immediately and fix the bug in the system. He eventually would come over, perhaps a week after the system had problems. As you can guess, the high opinion that the company had for the in-law was at an all-time low. To make things worse, the in-law was having problems with his wife and subsequently filed papers to divorce her. Needless to say, the company was left with a troublesome computer system, angry company officials, and disgruntled family members!

There are several morals to this story. First, many times it is more worthwhile to hire a company to put in a computer system and maintain it than it is to have a relative put the system together, part-time, during free-time, and at "no cost." Second, sometimes it is difficult to mix business and pleasure. In other words, you may not want to have your brother-in-law performing surgery on you. Likewise, you may not want your brother-in-law to set up your computer system, as possible family conflicts may set in. Third and last, make sure your company is trained on how to use the computer system. Documentation on how to use the system and software, as well as training sessions, should be provided to the users of the computer system. The most creative system will *not* be used if people don't know how to use it. And, a bad experience with computers (or in-laws) may turn a person against them!

In the Trenches #2: Learn the Lay of the LAN (Local Area Network)

This story involves the same company discussed above regarding installation of its computer system. Since the company wanted to have multiple users using the computer system at the same time within the company, it was wisely determined that a *local area network* (LAN) was needed to allow multiple users to access the computer. A LAN is a communication system linking two or more personal computers, allowing for sharing of software and data. A LAN is an important vehicle for dispersing and sharing information (including "electronic mail") within groups, as well as sharing expensive devices such as laser printers. About 80% of the information generated by an organizational group is used only within the group. This number substantiates the need for having a LAN so that a given group can exchange this information.

A major problem that the company experienced was that the LAN would "go down" frequently, which affected the sharing of software and information within the small company. If the network goes down, then this is a major problem, as it would be difficult to access the software for multiple use. In this company's situation, Murphy's Law applied, because not only was the company having specific problems with the software but also the company was having trouble with the network to even allow them to use this software. Who do you call? You guessed it — the brother-in-law. So again, assuming there was free time on the in-law's schedule, he would come in to fix the network problem. Of course, a log of the computer and network problems was not kept that would allow a record indicating how problems were solved. The in-law would come in, fix the problem, would not record the problem in a log, and then would leave to go to a "billable customer."

Again, the moral to this story is that a LAN is a good idea to use in your office if requirements necessitate its use. However, get someone to install it and maintain the network and computer system properly. Paying some money for this service, as opposed to *ad hoc* free help, will probably save you many hours of headaches and aggravation later.

In the Trenches #3: You Are Safer With a Backup

Imagine spending many hours on the computer, inputting important information into the computer on your patients' or customers' records, and then realizing that you either forgot to save this information in the computer, made a mistake, or there is a power failure and all your newly inputted information is lost! You think, and justifiably so, "What a waste of time!" What do you do? Well, there are some utility software packages that might allow you to recover some of this information, but generally speaking, you are probably in trouble.

This scenario is not far-fetched by any means. It has become standard practice for companies that use computers to backup their records daily, or perhaps weekly. *Backup* refers to the process of copying a file or an entire disk so that data are not lost when a disk is damaged. This process may be easily forgotten because people may think that a computer doesn't break or that the likelihood of a damaged disk or power outage is small. However, why take a risk if you don't necessarily have to take it? The bottom line to this story is to get in the practice of backing up your files either on disks or magnetic tapes (if warranted) so that you don't have to lose important information out of carelessness. Some systems, however, automatically save your programs and data periodically.

Summary

With these simple words of wisdom, you will save countless hours of possible mistakes. Heed this advice and it will help you get started in your "computerized office."

Putting the Cart Before the Horse at a Major University

Introduction

In today's competitive environment, universities are starting to embrace the concept of total quality management in order to enhance and refine their processes and functions toward providing better services, educational and otherwise, to the student. Some universities are slower to react to this increasingly competitive environment and will likely suffer financially due to lower student enrollment. One such university that needs to cater more to the needs of the student is located in a major metropolitan area. A few vignettes will show how information in this university is not being managed well in the student-oriented systems.

Vignette 1: The Grade Recording Process

Correctly recording the grades on a student's report card is a vitally important function of a university. If the grades are incorrectly recorded, then the educational stamp placed by the university will be severely damaged.

At one major university on the East Coast in the U.S., it has been rumored that up to 40% of the grades in a given semester may be incorrectly recorded on the student's report card and associated transcript. If this number is true, the university is faulty, irresponsible, and potentially liable for such errors.

The grade recording process at this university works as follows. At the end of each semester, the professor receives a computerized printout of the grade sheets, which has the alphabetized list of students in the professor's class and a column for the final course grade. The professor writes in the grades for each student and then gives the original grade sheets to the registrar's office. The registrar's office then has data entry clerks manually input the grades from the grade sheets into the computer system. At this point in the process, two observations can be made. First, the data clerks have mounds of grade sheets to enter into the system and some of these clerks are wearing headsets and not paying proper attention during this inputting process. Second, the grade sheet format does not match the format in the computer system used by the data clerks. Evidently, the grade sheets are alphabetized by name but the computer system format for the grade sheets is by student number. Thus, the data clerks have to hunt and peck on the professor's grade sheet for matching the final course grade for a student on the computerized system. Herein lies part of the problem. Once the grades are inputted, the report cards are sent thereafter to each student.

There are several ways to easily correct this grade recording problem. First, the data clerks could scan in the grades instead of having to manually input each grade. Second, the professor could even use an on-line system to input the grades directly into the computer. Third, if this is not possible, the format of the professor's grade sheets should match the format of the computerized version. Last, if all these changes are not possible, the manager of the grade input section should make sure that distractions while inputting grades (such as headsets, etc.) are kept to a minimum.

These recommendations seem so obvious, one might wonder how the computerized grade recording process could have been developed without considering these issues. Upon investigation into this problem, it was found that one of the chief users of this grade recording system (namely, the professor) was never asked for comments, suggestions, and feedback while the system was being designed. Incorporating the end user into the design, development, and implementation process is a critical step in systems development. The professors (or even a focus group of professors) were never consulted and involved in the system development process. This was a fundamental mistake in the system development process. As a result, inefficiencies and errors are being made.

Vignette 2: The Registration System

The next vignette shows another information-systems-related problem, due to management oversight, at the same university.

Besides grading, another critically important student-oriented system at a university is the registration system. A typical registration system will include the following steps:

- Student requests a class/course;
- A verification of the seats available is done by looking at the class schedule and the number of available seats;
- If seats are available, a class seat will be assigned following verification that course prerequisites have been met;
- The student transcript will be analyzed to see if the prerequisites for the requested course have been taken by looking at the catalog and comparing the courses required with the student's transcript;
- If the course prerequisites have been met and seats are available, the class request is accepted;
- The student's social security number is posted to the registration list and the scheduled class is then included in the student's tuition bill.

At the major university from the previous vignette, the registration system did the above steps except for one important function — the course prerequisites were never checked and this capability was never encoded into the university registration system. Without this check-and-balance, students are able to enroll in classes where they don't have the prerequisites. This puts the student at a disadvantage, as well as going against the professor's desires and university catalog's function. By not checking whether the student has the prerequisites, the class composition is affected in terms of the students having a similar, consistent academic background for taking the course. Again, a fundamental flaw in this registration system exists partly because of poor insight on the part of the system developers, and partly because of not asking for input from one of the main stakeholders — the faculty.

Vignette 3: No Back-Up System

Several poor academic and administrative decisions have been made at this major university under study. The two previous vignettes stressed the lack of end-user input and feedback into the grading and registration systems. Many times administrative decisions are made in isolation of academic interests. Another example of such a decision is not having a back-up plan for graduation ceremonies, in case of inclement weather. Any good (or even bad) information systems specialist will know to make back-ups and have contingency plans. In the case of this major university's outdoor graduation ceremony, it was canceled (and rightly so) due to lightning. There was never any contingency plan established in case of poor weather, and thus the graduates (many of whose family members flew in from abroad for graduation) never had the graduation ceremony as scheduled for that day. Of course, a ceremony was planned a few weeks later but this offer did not make up for the extremely bad planning on the part of the university's top administration. It certainly didn't help the families who spent thousands of dollars to attend their child's graduation for naught. And it certainly will not help the university's development office in trying to obtain donations and support from this graduating class' alumni. Again, this vignette shows how the disease of "bad judgment" spread from the university's systems development practices to even graduation.

Vignette 4: Expansion of the Curricula without Additional Resources

Another example of poor management acumen applied at the same university deals with approving the expansion and remodeling of a graduate curriculum without adding additional resources to cover this expansion. For example, in this revamped curriculum, there will be a core information systems course that every student will need to take (unless they waive out of the course via a waiver examination). This will add 10 new sections of this course every semester. To adequately cover these new course sections and others, three to four new faculty positions are needed. So far, the administration has not granted these new faculty positions to the department, even though the student enrollment is rocketing in the graduate information systems degrees. Also, additional computer facilities are needed to meet these needs and it is not clear if such resources will be forthcoming.

This lack of good judgment, along with those exhibited in the other vignettes, typifies poor planning on the part of the university administration. If this lack of concern for the students continues to permeate other systems and decisions affecting the students, then student goodwill will dramatically decrease, resulting in future decreased student enrollments. Again, the lesson learned here is to pay attention to the end users and stakeholders of the systems being developed and then make appropriate, well-informed decisions.

8 A Day at the Doctor's Office

Do I Really Need Automation?

Let me convey a real story to you regarding someone thinking about getting computers for the office.

Anecdote #1: Do It Right!

I know a doctor who was deciding to slowly phase computers into her office. One day, the doctor, before consulting others, saw that there was a great sale on a leading-brand electric typewriter and accompanying computer. The

It's Time to Think About Automation!

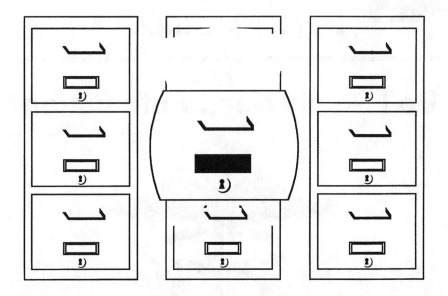

doctor probably felt that since this company made great typewriters, then they should also make very good computers. So, after buying the typewriter and computer, the doctor asked a friend of hers to set up the typewriter and computer so that several types of letters could be stored in the computer and then the receptionist could change the "form" letters according to the doctor's recommendations for a given patient. There were about seven standard types of letters that the doctor usually sends. Among these were a patient referral letter, an insurance type of letter, even a "please go elsewhere" letter (if the patient is uncooperative for one reason or another), and several other types of letters. Essentially, the doctor wanted to use the computer for limited word processing capabilities.

Even though the doctor had good intent, there were several mistakes that the doctor made that resulted in the receptionist never using the computer. The first mistake was that the doctor did not look to see what her colleagues had in their offices for a computer system. The doctor did not seek out the advice of other doctors to learn from their successes and mistakes. The doctor could have also asked "someone in computers" to identify the requirements needed for a computer system in her office, and then get advice on what hardware (equipment) and software (computer programs) are recommended for her needs. As a result of not doing this, the doctor had an awkward typewriter-computer system in which the keyboard and a separate keypad both had to be used to update information and then print out letters. This was such an awkward system that it would have been almost easier to simply retype each letter instead of retrieving a letter from the computer's memory and then altering it. Another problem with this system was that it could not be easily expanded in order to satisfy future needs of the doctor — such as including a payroll, billing, and/or an appointment scheduling function.

The second major mistake was that the doctor spent money foolishly for a system that would not be used for its intended purposes (i.e., the computer would not be used at all; only the typewriter would be used for typing letters). The doctor should have spent the time either by herself, or through a medical computer consultant, to carefully analyze her requirements for using a computer and then to carefully determine the alternatives for setting up the computer system *and* training the office staff on how to use it. The doctor tried to force-fit her requirements to the computer, instead of trying to determine which computer system would satisfy her requirements.

The third and last major mistake was that, by having such an awkward computer system, the receptionist became disillusioned with the computer (and possibly all computers after this event), and this created uneasy feelings between the doctor and the receptionist. Here the doctor spent the extra money for this computer, and the computer was just collecting dust. If the doctor had realized how difficult it was to use this particular computer system, the doctor would have quickly recognized the big mistake that she made.

Anecdote #1: Lessons Learned to Help You Not Make the Same Mistakes

Lesson 1:
Do your homework or have someone else do it for you.

Spend the time up front learning from others what they have done in setting up their computer system for a practice/business similar to yours. Besides checking with your colleagues, you should also consult popular computer magazines, such as *PC* or *BYTE* (or now, the Web), to find general information on computers. Doing this research at the beginning will prevent you headaches down the road. If you can't or don't have the desire to do this research by yourself, then obtain a computer consultant to help you with these decisions. Your local professional association should probably have a list of recommended computer consultants in your area. Also, you might want to attend a computer conference to find out about the latest technology and benefits to you for your business. You might check with your national association, such as the American Medical Association or American Bar Association, about their listing of conferences, which typically include sessions or seminars on computer-related topics.

Lesson 2:
Don't put the cart before the horse

Many times people buy computers or install a computer system and then decide how the computer system could be used to help them in their business. Or they buy software and then see if it can be used in their business. The problem here is the lack of determining requirements *before* buying the computer system. Time invested in the requirements definition stage will save you time and costs when you install and implement your computer system in your office.

Lesson 3:
Don't introduce something new and then leave town

This lesson means that you shouldn't introduce a new technology into your business without providing proper training to those people in your business who are affected by this change. You should keep the staff abreast of the changes during the acquisition and installment of the computer system. Additionally, training should be provided to those using the computer in order to make them feel at ease and comfortable when working with the system. Also, proper documentation describing the hardware and software should be provided by the computer/software vendors in case there are questions.

Lesson 4:
Just because you have a computer, it doesn't mean that it can't break!

Just as with everything else, you need a backup plan; with computers, it is no different. There might be errors in the program, there might be a computer hardware problem, or the power might be off in your office so you can't use the computer or a list of other causes. For these reasons, you should take proper measures to guard against some of these problems. One thing to do is to have a backup system in place, perhaps a manual one like the way you were doing business before having the computer. Also, you should "backup" your files daily or weekly to guard against loss of damaged data or records. The idea of a recovery plan is important in knowing how to do business until the computer functions properly again and in knowing how to restart operations when the computer is back. This concept of "backing up" will be explained in a later chapter. Just when you least expect it, Murphy's Law sets in and the wheels of your business will still need to continue to turn. That's why you need to *plan ahead.*

MANAGEMENT, DEVELOPMENT, AND LEGAL CONSIDERATIONS

9 Understanding the Corporate Culture

Understanding Culture is Critical!

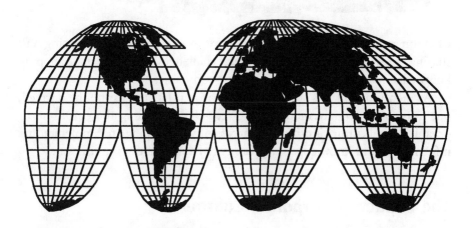

In a recent knowledge management study conducted for The World Bank in Washington, D.C., one of the major findings was, "In all the organizations studied, culture was one of the key factors either enabling or inhibiting effective knowledge sharing." Whether competitive or cooperative, the knowledge management system in an organization needs to be consistent with the culture. Organizations, such as Lotus Development, McKinsey, and

Build From Your Successes and Failures!

Chevron report that changing the culture to encourage documenting and sharing knowledge was one of the biggest challenges they faced in developing knowledge management systems in their companies.

As mentioned in this study, culture change is hard. It cannot be decreed, but must be genuinely modeled and led. Understanding the corporate or organizational culture is critical in successfully developing and implementing information and knowledge management systems in an organization.

The Impact of Corporate Culture

There are numerous evidences of culture in an organization. According to Marilyn Parker in her book *Strategic Transformation and Information Technology,* some examples of culture are the greeter at the Wal-Mart door, the conservatively dressed IBM salesperson, or the casually dressed university professor. In Kotter and Heskett's study on corporate culture and performance, they concluded that:

- Corporate culture has a significant impact on a firm's long-term economic performance;
- Corporate culture will be an increasing factor in determining the success/failure of firms in the next decade;
- Management can force corporate culture to become more performance enhancing, although it may be difficult to do.

The importance of corporate culture is closely linked with the success (or failure) of information technology projects within the organization. If the information system being proposed is not consistent with the organization's culture, then acceptance of the project may be difficult. Additionally, if users feel uncomfortable with the addition of the information system as part of their routine functions, then resistance will heighten and the system will not be utilized.

According to Parker, unhealthy corporate cultures exist where:

- Managers place a low value on the opinions and wishes of customers and stockholders;
- Managers behave politically;
- Managers place a low value on leadership and on the employees who can provide it;
- Managers tend to stifle initiative and innovation and behave in centralized and bureaucratic ways.

She further elaborates that creating a new performance-enhancing culture requires the following steps:

1. Leadership from the top (excellent leaders with an outsider's broad perspective and an insider's credibility);
2. These leaders must convince and motivate others about a new vision and a new set of strategies for the firm;
3. From this, a new corporate culture emerges. A growing coalition of managers begins to share the values of top management;
4. Behavior and practices change;
5. Leaders must continue to communicate about core values and behaviors to preserve the culture.

It can take from 5–15 years, according to Parker, to create a new culture and preserve it. Stifling corporate hierarchies, which cause inflexibility and failure, contribute to an unhealthy corporate culture. The need for thinking, learning organizations is critical in order for today's organizations to meet and cope with the challenges of tomorrow.

Suggestions on Dealing With Organizational Culture and Information Systems

Various researchers have studied IS technology implementation and organizational culture. Pliskin, Romm, Lee, and Weber label "dimensions" of culture that should be considered prior to implementing new technologies. These include reward orientation, integration/lateral interdependence, innovation and action orientation, decision-making autonomy, and performance orientation. There is a strong need for alignment between the actual culture in a company and the cultural assumptions embedded in the system by its designers. Pliskin and colleagues feel that an MIS (management information system) must be designed and adjusted to fit the organizational culture.

Michael Gallivan's work at New York University emphasizes that managers should be strongly advised to evaluate the potential implications of a new technology before adopting it — by conducting a cultural analysis. Michael Ginzberg at Case Western suggested a similar approach earlier by allowing a broad cross-section of employees, users, and managers to evaluate their reactions to the system early on (to establish "early warning indicators"). Janice Burn of Edith Cowan University in Australia suggested a research framework for taking into account cross-cultural factors relating to IS development and implementation. Her framework includes performing a societal culture analysis, organizational analysis, and information culture analysis.

Achieving Success With Matching IS to Culture

One of the most important lessons for any IS developer is to pay close attention to the intraorganizational dynamics and politics within the organization. A first giveaway is the formal organization chart of the company. If there isn't a CIO (chief information officer) or VP-IS or if the IS director reports to the VP-Finance, this gives an automatic clue to the culture in the organization and the role and respect of IS in the organization. In one company, the IS director reported to the VP-Finance, which is the "old conservative style" of reporting. This immediately shows that IS doesn't have the esteemed respect and authority that other parts of the company command (i.e., IS plays a subservient role instead of a leading one in the company).

Another indicator of corporate culture and its impact on IS can be given by the investment in IS hardware and software in the company. If one walks through (i.e., "management by walking") the company, what types and models of computers are on the desks of the employees and management? What kinds of databases and groupware are being used? How archaic are the information systems? Simply observing what is happening in the company will give the IS developer a feeling for the organizational climate and the role that IS plays in it.

The IS developer should not get bogged down in the "technology" side of the equation. Rather, he/she should pay careful attention to some of the management indicators such as user involvement and enthusiasm, top level support, aligning IS with the strategic goals of the company and to its bottom line, and other important factors to ensure a greater likelihood for project success. Marketing the value of the IS being developed or implemented needs to be a continual effort to further increase awareness and use of the system.

With these few tips in mind, the IS project will likely fit within the organization's culture and allow the system to be successful.

10 Have a Good Follow-Through

Just remember--the key is in the follow-through!

Implementation is a critical part of the system development life cycle that is often overlooked. Many implementation problems involve either not considering implementation issues during the planning stages of the project or a poor transition or hand-off is made from the developers to the implementers of the system (i.e., the "over-the-wall" phenomenon where the system developers throw their system over the wall to the integration and testing group, who then throw the system over the wall to the users). As in golf or tennis,

Know Which Way You Are Heading!

if the follow through of the club or racket is not done properly, your shot will be less than desired. Similarly, if implementation and institutionalization are not adequately addressed, then your system will most likely lead to failure.

John Stone, Chief Technology Officer at the world's largest financial printer, Bowne & Company, identifies six critical success factors for application development in a multiple and evolving information technology environment. In his 1997 book titled *Developing Software Applications in a Changing IT Environment,* Stone describes the six factors as follows:

- Meaningful project reporting;
- Appropriate organization and culture;
- Appropriate support infrastructure;
- Accurate estimation and scope control;
- Correctly staffed team;
- Sufficient business user involvement.

Barbara Bouldin, in her book, *Agents of Change: Managing the Introduction of Automated Tools,* devotes several chapters just to implementation. Some of her implementation guidelines follow:

- Divide the implementation effort into manageable chunks;
- Manage change implementation via incremental improvement;
- Promote a comfortable situation for people by being flexible and meeting each user on his/her own ground;
- Expect numerous unforeseen problems;
- Always expect new users to train and new managers to sell;
- Develop a user guide that will document aspects of the change that are particular to your specific environment;
- Form a local user group as an effective means to keep your change process vital;
- Following a reorganization, expect to recreate the proper environment by selling your new management and supervising and training your new group;
- Measure the benefit of the system by progress reports with percentage completed information, publication of significant and dramatic numbers, analysis and graphic depiction of trends, development and analysis of user satisfaction surveys, comparison of original projections and estimations to actual statistics, and utilization of tools to measure productivity in terms of quality and reliability.

According to Foundation Technologies in Cambridge, Massachusetts, there are critical steps for achieving information systems (and specifically, expert systems) success. These steps include forming coalitions with the full range of stakeholders, educating all levels of the organization, working with the MIS group, tapping into and becoming part of the existing business visions, and working to achieve and sustain high level executive support. To establish sustainable value, stay in line with the business vision, provide sufficient speed (the system operates in "real" time), allow for integration, plan for changes in behavior (e.g., middle management education), and understand the capacities of the technologies.

Terry Byrd, Richard Will, Richard Hauser, Laura Davis, and others have written papers looking at implementation issues, primarily related to expert systems (however, they could apply to most information technologies). Byrd points out that the lengthy implementation time and job impact (i.e., stress and loss of control) are two important factors affecting successful implementation of expert systems. Richard Will identifies the two most important issues for long-term expert system success as maintenance and verification and validation. Richard Hauser, in his work on managerial issues in expert system implementation, believes that performance appraisal systems should be redesigned to include the expert system performance as a part of the employee's evaluation. This step would allow employees to accept ownership and take responsibility for the performance of the expert system. Laura Davis, at the Navy Center for Applied Research in Artificial Intelligence, discusses transitioning expert system technology. She states that it is very important to dedicate personnel to technology transition activities. Problems that arise once systems exchange hands can be particularly time-consuming in the areas of porting to new hardware, explaining and supplementing documentation, and training personnel through formal demonstrations or informally responding to queries. She found that success of AI depends on a cultural change that must go hand in hand with the technology implementation.

In Peter Duchessi and Bob O'Keefe's paper titled "Understanding Expert System Success and Failure" in the *Expert Systems With Applications* international journal, they studied six expert system projects at IBM, GE, and Kodak. Three were successful implementations and three were not too successful. The applications ranged from financial to engineering tasks. Duchessi and O'Keefe discovered the following: top management support and immediate manager acceptance are important; demonstrable benefits and problem urgency affect management support; and at the user level, perception of management support, degree of organizational change, organizational support, and users' personal stakes in the system affect operational use.

A key point to remember in the development and implementation stages of an information technology product is that no matter how well the system performs, it won't do any good if no one can or will use it. You may have a technical success but a technology transfer failure. Some general guidelines for proper implementation include the following:

- Know the corporate culture in which the system will be deployed;
- Planning for the institutionalization process must be thought out well in advance, as early as the requirements analysis stage;
- Through user training, help desks, good documentation, hotlines, etc., the manager can provide mechanisms to reduce "resistance to change";

- Solicit and incorporate users' comments during the analysis, design, development, and implementation stages of the system;
- Make sure there is a team/individual empowered to maintain the system;
- Be cognizant of possible legal problems resulting from the use and misuse of the system;
- During the planning stages, determine how the system will be distributed;
- Keep the company's awareness of the system/technology at a high level throughout the system's development and implementation, and even after its institutionalization.

11 Integrating and Evaluating Decision Support Systems into the Organization

Teamwork and Collaboration

In order for decision support systems (DSS) to be properly designed, implemented, and used, the user must be incorporated into the DSS development process. The user's comments and feedback should be part of the DSS building process. However, even though the user is involved in the design of the

It's a Deal!

DSS, it is equally important that the user be involved in the post-audit evaluation of the DSS. The evaluation process entails two major areas. One area is to make sure that the DSS is easy to use and accurate. The second area is to check that the DSS is designed to meet its intended objectives. Once the DSS evaluation is performed, then the DSS can be integrated into the organization. This chapter first discusses DSS evaluation and then examines various integration strategies.

Evaluation of DSS

Over the years, there have been several general frameworks proposed for evaluating DSSs. Boehm et al.[1] identify several important characteristics of quality software:

Portability

- Device-independent: software can be executed on computer hardware configurations other than its current one.

Reliability

- Completeness: all of the software parts are present and each of its parts is fully developed.
- Accuracy: its outputs are sufficiently precise to satisfy their intended use.
- Consistency: it contains uniform notation, terminology, and symbology within itself, and its content is traceable to the requirements.

Efficiency

- Device efficiency: it fulfills its purpose without waste of resources.
- Accessibility: it facilitates the selective use of its components.

Human Engineering

- Communicativeness: it facilitates the specification of inputs and provides outputs whose form and content are easy to assimilate and are useful.

Testability

- Structuredness: it possesses a definite pattern of organization of its independent parts.

Understandability

- Self-descriptiveness: it contains enough information for a reader to determine its objectives, assumptions, constraints, inputs, outputs, components, and status.
- Conciseness: no excessive information is present.
- Legibility: its function and those of its component statements are easily discerned by reading the code.

Modifiability

- Augmentability: it easily accommodates expansions in data storage requirements or component computational functions.

Gaschnig et al.[2] identify evaluation characteristics for software, specifically expert systems. These characteristics are the quality of the system's decisions and advice, correctness of the reasoning techniques used, quality of the human–computer interaction (both its content and the mechanical issues involved), the system's efficiency, and cost-effectiveness. User-oriented evaluation criteria for DSS software are important elements of the DSS evaluation process.

According to Rouse,[3] there are three primary evaluation concerns with decision support systems: compatibility, understandability, and effectiveness. A decision support system is compatible if the nature of physical presentations to the user and the responses expected from the user are compatible with human input–output abilities and limitations. A DSS is understandable in the sense that the structure, format, and content of the user–system dialogue result in meaningful communication. A DSS is effective to the extent that it supports a decision maker in a manner that leads to improved performance, results in a difficult task being less difficult, or enables accomplishing a task that could not otherwise be accomplished.[3]

Rouse explains that there are two different approaches to evaluation in order to meet these three concerns. An empirical evaluation could be performed if a prototype system and population of potential users are available and if time and resources allow. This is a bottom-up approach that involves first assuring compatibility (via paper evaluation, e.g., checklists) and then assessing understandability and effectiveness (via part-task and full-scope simulator evaluations, respectively). A part-task simulator is a device that roughly approximates the real system of interest in terms of appearance, static and dynamic characteristics, and range of decision-maker activities required. A full-scope simulator is a high-fidelity replica of a real system that allows decision makers to experience virtually the full range of system behaviors, without waiting for all these behaviors to occur in the normal course of events or endangering anyone by initiating the situations of interest in the real system.[3]

An analytical evaluation could be used if the system only exists in terms of design documentation, or the population of potential users is not yet available, or time and resources are constrained.[3] The analytical evaluation is a top-down manner that involves viewing a DSS as if it were designed using the design process of execution and monitoring situation assessment: information seeking, situation assessment: explanation, and planning and commitment.[3]

Besides these evaluation methodologies, other frameworks exist to evaluate DSS. Adelman and Donnell[4] suggest that if an evaluation is to be effective, the evaluator must decide in advance what is to be examined. This is done by identifying one or many measures of effectiveness.[4] These measures of effectiveness relate to three types of interfaces — (1) DSS/user interface, (2) user-DSS/organization, and (3) organization/environment. A hierarchy of these measures of effectiveness upon each interface is shown below:[4]

DSS/User Interface

Match with personnel
    Training and technical background
    Workstyle, workload, and interest
    Operational needs
DSS's characteristics
    General
        Ease of use
        Understanding
        Ease of training
        Response time
    Specific
        User interface
        Data files
        Expert judgments
        Ability to modify judgments
        Automatic calculations
        Graphs
        Printouts
        Text

User-DSS/Organization

Efficiency factors
    Time
        Task accomplishment
        Data management
        Set-up requirements

Perceived reliability under average working conditions
    Skill availability
    Hardware availability
Match with organizational factors
    Effect on organizational procedures and structure
    Effect on other people's positions in the organization
        Political acceptability
        Other people's workload
    Effect on information flow
    Side effects
        Value in performing other tasks
        Training value
Organization/Environment
Decision accuracy
Match between DSS's technical approach and problem's requirements
Decision process quality
    Quality of framework for incorporating judgment
    Range of alternatives
    Range of objectives
    Weighing of consequences of alternatives
    Assessment of consequences of alternatives
    Reexamination of decision making process
    Use of information
    Consideration of implementation and contingency plans
    Effect on group discussions
    Effect on decision maker's confidence

In order to obtain values for these measures, several collection procedures could be used. The first procedure is subjective judgment, which requires users to score their experiences, usually by answering a questionnaire following use of the DSS.[4] Another technique is expert observation, which involves subjective judgment, only this time on the part of nonparticipating observers of the DSS users and experts in the area the DSS was developed to support. The last collection technique is objective measurement, which is usually associated with empirical experimentation and comparison.[4]

There are still other techniques for evaluating DSS. Expert Choice is one technique that could be used for evaluation. Liebowitz[5] shows how this process works for evaluating expert systems. Holzer[6] describes another evaluation approach that was used at Standard Oil of Ohio. Holzer uses a three-phase process for evaluating and selecting DSS software. The first phase is screening, in which DSS products are matched against a written checklist for required and desired capabilities. The next phase consists of an examination to check for capabilities and features of the DSS software. A detailed review of these DSSs is conducted, with a special focus on problem-solving power, ease of use (for non-systems users), and quality of vendor technical support. The last phase of the evaluation process involves user involvement and hands-on testing. Here, user reactions are collected via a detailed survey, and time is spent on learning the product, coding and running the benchmarks, and assessing quality of technical support.

Davis and Olson[7] indicate another approach for evaluating software effectiveness. They feel that there are several ways for evaluating system value. These include significant task relevance, willingness to pay, system usage, and user information satisfaction.[7] For a decision support system, task relevance is improved decision quality, which is often difficult to observe but sometimes possible to approximate through users' subjective estimates. Under willingness to pay, users may be asked to specify how much they are willing to pay for a specific report or system capability. Under system usage, system logs may permit measures of system use, or users may be asked to estimate their use of the system. With user information satisfaction, users are asked to rate their satisfaction with such aspects of the system as response time, turnaround time, vendor support, accuracy, timeliness, format of outputs, and confidence in the system.[7]

Last, Keen and Morton[8] suggest eight methodologies for DSS evaluation. These include the following:[8]

- Decision outputs: evaluate the system to see if the DSS led to improved actual decision results.
- Changes in the decision process: look at changes in the way decisions are made.
- Changes in managers' concepts of the decision process: see if the DSS helped change attitudes such as those toward the users' tasks and the use of analytic tools and toward other units of the organization.
- Procedural changes: evaluate if the DSS led to changes in those activities within the decision process that are in some sense phsycial, such as the mechanical procedures and the use of such resources as people, machines, and paper.
- Cost/benefit analysis: evaluate a DSS in terms of classical cost/benefit analysis.

- Service measures: evaluate a DSS in terms of responsiveness of the system, availability and convenience of access, reliability, and quality of system support, such as documentation and training.
- Managers' assessment of the system's value: ask managers what the DSS is "worth."
- Anecdotal evidence: supplement the formal evaluation process with insights, examples, opinions, and events collected by a skilled observer.

No matter what approach one takes to evaluate DSSs, it is vital to get the users involved in the process. This is critical to the evaluation process, as well as ensuring a smooth integration of the DSS into the organization.

The next section looks at implementation issues for introducing DSS into an organization.

Implementation Strategies

Even though a DSS is validated and evaluated, there is still no guarantee that the DSS will be accepted by users and management into the organization. This suggests that implementation plays an important role for DSS integration into the organization. In fact, implementation of a computer-based information system almost invariably involves change on the part of users and the organization as a whole.[9] Failure to institutionalize change has been one of the major pitfalls of implementation projects, particularly in the case of DSS.[8,9]

According to Keen et al.,[8] there are five major factors that influence the successful implementation of DSSs. These are top management support, a clear felt need by the client, an immediate, visible problem to work on, early commitment by the user and conscious staff involvement, and a well-institutionalized operations research/management science or management information systems group.

Without the financial and moral commitment from top management, the implementation of a DSS, and expert systems as well, would be very difficult. Since DSSs are typically developed to aid a manager or executive, top management needs to accept the development and use of a DSS so that policies could be made to encourage management in the firm to use the DSS. Of course, before top management can give their approval, the DSS should be designed for problems that are "causing a large amount of grief to a large number of people." This shows that the DSS is addressing a real need, and the DSS could supplement the decision maker's ability in solving the problem.

Besides the need for addressing a visible and important problem, successful DSS implementation entails getting the users involved during the DSS development, testing, evaluation, and implementation stages. User feedback should be incorporated into the design and implementation of the DSS. For example, the input/output content of the DSS (i.e., how the output should be displayed, the type of user interface, etc.) and training procedures on how to operate the DSS (i.e., training courses, user documentation, etc.) are important components for successful DSS implementation. Last, a well-seasoned and respected OR/MS or MIS group could help instill confidence in the users for operating the DSS.

Alter[10] feels that there are various risk factors that could affect DSS implementation. The risk factors are not unique to DSS implementation, but are common to other software implementations, such as expert systems. These risk factors, problems, and typical situations are shown below:[10]

Risk Factor: Nonexistent or unwilling users
Problem: Lack of commitment to use the system
Typical Situation: System not initiated by potential users and developed without their participation.

Risk Factor: Multiple users or implementers
Problem: Communication problems
Typical Situation: System involving voluntary use by many individuals or coordination among many people.

Risk Factor: Disappearing users, implementers, or maintainers
Problem: No one available to use or modify the system
Typical Situation: System is the vehicle of a person who leaves or system initiator leaves before system is installed.

Risk Factor: Inability to specify purpose or usage pattern
Problem: Overoptimism on part of system designer and advocates
Typical Situation: Assumption that noncomputer personnel will figure out how to use the system.

Risk Factor: Inability to predict and cushion impact
Problem: Lack of motivation to work or change work pattern without receiving benefits
Typical Situation: Forced changes in organizational procedures.

Risk Factor: Loss or lack of support
Problem: Requirements for funding; obstruction by uncooperative people
Typical Situation: Lack of budget to run system; lack of management action to use system effectively

Risk Factor: Lack of experience with similar systems
Problem: Unfamiliarity leading to mistakes
Typical Situation: Developing an innovative system aimed at substantive change rather than automation per se.

Risk Factor: Technical problems and cost-effectiveness
Problem: Cost of maintaining or improving system
Typical Situation: No adequate way to estimate value of system either before or after potential improvements.

To minimize these risk factors, there are various implementation strategies[10,11] to use. One strategy is to divide the project into manageable pieces. This involves using prototypes, using an evolutionary approach, and developing a series of tools. This will minimize the risk of producing a massive system that doesn't work. A second strategy is to keep the solution simple. To do this, be simple, hide complexity, and avoid change. This will tend to encourage use and to avoid scaring away users. A third strategy is to develop a satisfactory support base. To accomplish this, obtain user participation and commitment, obtain management support, and sell the system. The last strategy is to meet user needs and institutionalize the system. Providing training, good user documentation, and ongoing assistance can ensure having a DSS for many individual users in an ongoing application. Permitting voluntary use and tailoring the system to people's capabilities are important components of this implementation strategy.

It is not essential to use only one of these strategies. In fact, most circumstances would require using all of these strategies or combinations of them. When developing the system and implementing it, the designers and implementers should keep three factors in mind:[8]

1. Have mutual commitment and realistic expectations by designers, sponsors, and users;
2. Recognize and resolve resistance to change;
3. Institutionalize the completed system.

Overexpectations are sometimes a big dilemma with DSS and ES development. In the expert systems area, the hype of this new technology by the developers and expert systems companies created overexpectations. Likewise, in the DSS area, realistic expectations must be made as to the capabilities and limitations of the system. Resistance to change can be ameliorated if the user is involved in the process, and if the system can be developed to make the user's job easier. Incorporating the comments of the users is vital to successful implementation. Institutionalizing the completed system involves providing proper training and assistance, and having top management encourage use of the completed system to make it one of the company's necessary decision aiding tools. Designating individuals to maintain and update the system, especially in the case of expert systems, is critical to the ongoing use of the system.

If these strategies are followed and these risk factors are recognized up front in the development process, then the DSS implementation should be a smooth and harmonious process.

References

1. Boehm, B., J. R. Brown, H. Kaspar, M. Lipow, G. J. MacLeod, and M. J. Merrit (1978), *Characteristics of Software Quality*, North-Holland, Amsterdam.
2. Gaschnig, J., P. Klahr, H. Pople, E. Shortliffe, and A. Terry (1983), "Evaluation of Expert Systems: Issues and Case Studies," in *Building Expert Systems*, F. Hayes-Roth, D. A. Waterman, and D. B. Lenat (Eds.), Addison-Wesley, Reading, MA.
3. Rouse, W. B. (1986), "Design and Evaluation of Computer-Based Decision Support Systems," *Microcomputer Decision Support Systems: Design, Implementation, and Evaluation*, S. J. Andriole (Ed.), QED Information Sciences, Wellesley, MA.
4. Adelman, L. and M. L. Donnell (1986), "Evaluating Decision Support Systems: A General Framework and Case Study," *Microcomputer Decision Support Systems: Design, Implementation, and Evaluation*, S. J. Andriole (Ed.), QED Information Sciences, Wellesley, MA.
5. Liebowitz, J. (1986), "Useful Approach for Evaluating Expert Systems," *Expert Systems*, Vol. 3, No. 2, Learned Information, Oxford, England.
6. Holzer, D. C. (1985), "Selecting a DSS Generator: A Case Study," *Proceedings of Decision Support and Expert Systems Conference*, George Washington University/U.S. Professional Development Institute, Washington, D.C., April 22-25.

7. Davis, G. B. and M. H. Olson (1985), *Management Information Systems: Conceptual Foundations, Structure, and Development*, McGraw-Hill, New York.

8. Keen, P. G. W. and M. S. Scott Morton (1978), *Decision Support Systems: An Organizational Perspective*, Addison-Wesley, Reading, MA.

10. Alter, S. L. (1980), *Decision Support Systems: Current Practice and Continuing Challenges*, Addison-Wesley, Reading, MA.

11. Reimann, B. C. and A. D. Waren (1985), "User-Oriented Criteria for the Selection of DSS Software," *Communications of the ACM*, Association for Computing Machinery, Vol. 28, No. 2.

12. DeSanctis, G. and J. F. Courtney (1983), "Toward Friendly User MIS Implementation," *Communications of the ACM*, Association for Computing Machinery, Vol. 26, No. 10.

12 Avoid Being the Political Football

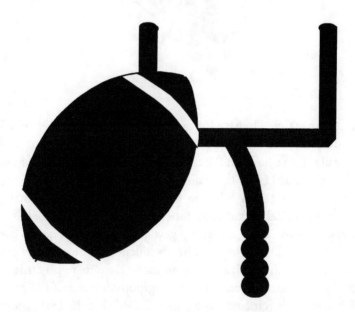

Have you ever felt that you may be the football being passed around by two opposing teams? This situation recently happened to me while working on a NASA application to develop an autonomous satellite ground control center. My task was to determine the most appropriate expert system tool that could be integrated into an existing NASA language for use by the flight operations personnel.

Three days into the project, I noticed that there was another group, assigned by the same group who oversaw my work, tasked with the exact same assignment. What was happening here? After doing some investigative work, I discovered that, according to the group for whom I was working, the

Lights-Out Automation is Fast Becoming a Necessity!

other group claimed to have the necessary artificial intelligence (AI) expertise to accomplish their assignment. But when I spoke with their lead, they said they really didn't have an AI expert on their team. Also, this group was in another Code at NASA that was a rival to the Code where I was working. My colleague at NASA told me that he didn't trust the other group, but because of political pressures and the hope of producing a better relationship between the two Codes, my group gave them a little money to carry out this task. Thus, I was brought in to be an independent evaluator, being thrown in the middle of this political football game. If I were to play this role, I would certainly antagonize at least one of the groups; it was as if I were being set up.

What was the remedy? I quickly surmised that the best approach for me was to collaborate with the other group in serving as the AI expert on their team. Instead of being an independent evaluator, I would rather serve as the AI expert member of their team since they were tasked to do the same job as me and they lacked the AI expertise.

Now the next question was, would my group allow me to play this role, since they were probably hoping that the other group would make mistakes due to their lack of knowledge in the AI area. In this manner, my group could outshine them and say "I told you so."

The other question was whether the other group would want me, as they may be suspicious from the start as to why I was tasked with the same assignment as theirs — would I be a "plant," an informant to my group by being on their team? There was an initial lack of trust, but they also realized they needed my help in order to produce hopefully successful results.

Another interesting political observation that I made was that my NASA colleague (who I was working for and who handled the management and budgetary aspects) seemed to have disagreements with his "technical" counterpart on the team. For example, the programmer who my colleague told me to work with was quickly usurped from this role 1 hour after the technical counterpart learned of this situation. This also seemed strange, as the programmer hadn't been assigned work in 2 months by the technical manager. Thus, there appeared to be political battles all around me.

At the time of these events, the NASA organization was in the midst of reengineering, restructuring, and downsizing. Almost each week, a new proposal for a reengineering structure was introduced (e.g., project-based teams and branches, discipline-based teams and branches, etc.). There seemed to be some low morale problems, as there was instability in the system.

So What Happened?

After speaking with my NASA colleague (who essentially controlled the money and was higher ranked than his "technical" colleague), I was given permission to work with the team of individuals who needed an AI specialist. The team, in the end, was also happy to have me as a team member, because they needed the AI talent to succeed in accomplishing their task. It turned out to be a win–win combination, although some people wanted our group to fail due to some in-fighting among the Codes. We selected the expert system shell and the AI system was built as promised, which contributed toward building an autonomous satellite ground control center. This new control center would become a model for others to follow.

Lessons Learned

1. If you can't beat them, join them.
2. Be careful of being thrown to the lions.
3. Avoid being caught in the middle of political battles.
4. It is difficult to get the technical work done without being mindful of the possible management and organizational barriers.
5. Don't necessarily try to outshine others — be a team player.

13 How Tight a Grip is Necessary?

Under the Thumb

Have you ever been engaged in a project where there is too much structure? Usually projects may not have enough structure and proper management. However, the following is a story that relates to an expert systems project that seemed almost too rigid.

The story involves a company that was tasked to develop an expert system prototype to act as a performance support system for helping engineering designers. The scope of the domain was well-bounded and more specific than "engineering design," but for confidentiality reasons, I can't disclose the exact application area. The project involves the cooperation of three companies and a consultant, all of which would develop various aspects of the knowledge-based system. Project management was very action-oriented, with deliverables, milestones, and frequent briefings and meetings. Each meeting was rather formal with an hourly agenda, transparencies, and audio–visual aids. It appeared that everything had to be planned out in advance and be well structured. This orientation was perhaps due to the engineering mindset, as almost everyone on the project had a traditional engineering background.

The consultant and one of the other team members had an information systems (IS) and knowledge engineering education. Their orientation was not so rigid that everything had to be planned out in advance. One major difference between their philosophy and that of the engineers was that the IS designers believed that a prototype expert system would ultimately be built, with some throw-away versions of the prototype early on until the system properly modeled our domain expert's reasoning process and knowledge. The more traditional engineers felt that throw-away versions of the system could not be tolerated and everything had to be done correctly the first time.

Another difference was that the engineers (not the *knowledge* engineers) on the project wanted all the questions to be asked of the expert in all the knowledge acquisition sessions be determined in advance. The consultant and the IS designer knew that the domain had some ill-structuredness, so uncertainty played a factor in this application area. It was a good idea to write down questions in advance of a knowledge acquisition session to be used as a guide throughout the interview, but an exhaustive list of all possible questions would be difficult to assemble. This is because the expert typically would bring new information and insight during a knowledge elicitation session which may not have been anticipated and cause new questions to be created and would greatly influence the pertinence of which "listed" questions to ask.

One other major distinction was that the traditional engineers were almost forcing the expert to model his way of approaching a problem to match their structure. It should have been the other way around, where you listen to how the expert is reasoning and using his knowledge, and then you structure and represent the knowledge accordingly. Fortunately, the consultant who acted as the knowledge acquisition facilitator prevailed in doing things the right way!

What can be learned from this experience? Two important lessons come to mind. First, it is perfectly fine to have a "game plan" in mind when developing a knowledge-based system. In fact, you need one, but it should allow for some flexibility in your approach and schedule. If it is too rigid, you may not be able to adapt easily to change. The field of knowledge acquisition, in particular, is as much an art as a science — probably even more so.

Second, it is necessary to take the lead in situations that involve your expertise, while others outside your specialty area may try to influence and perhaps sidetrack you. They may feel that their way is correct, even though they are not very knowledgeable or expert in the area. Don't succumb to their pressure, especially if you feel their approach is incorrect and outside their scope of talent. If you do not voice your opinion, the project may go down the wrong path.

14 Use It or Lose It

I'm Swamped! Too Many Things to Do in Too Little Time!

In many organizations, whether universities, government, or industry, a common practice is to make sure that you use up the money that has been allocated to you before the end of the fiscal year. If you don't exhaust each dollar in your budget, your budget may be reduced the following year, because not depleting your budget shows you don't need as many funds for next year's budget. Thus, for department managers, branch heads, chiefs, or other managers, there typically is a flurry of procurement activity during the fourth quarter of the current year in order to avoid a possible reduction in funds for next year's budget.

Anything you can do, I can do better!

This is a golden opportunity for contractors and consultants to be hired, and for employees to request use of these funds for research and development, hardware/software acquisition, travel, and the like (of course, within the limitations of the "color" of the money).

I became involved in an expert systems project within the last few months of the fiscal year simply by discovering that there was some extra monies available and the branch head was eager to spend this money to jump-start a research project for possible full funding next year. From my perspective as the consultant/contractor, this was essentially a newly found opportunity that fit my interests professionally, technically, and fiscally. This also opened the door to help develop the proof-of-concept idea and to then become involved on the research team if the project were to be fully funded.

From the manager's perspective, the usage of funds could be justified and the genesis of the project could commence. The manager needed someone with expertise in the expert systems and knowledge acquisition area, and I felt fortunate to be able to fill the bill! The project was quite interesting from an applied research perspective, so it appeared to be a win–win situation for both the manager and myself.

What lessons learned could be learned on the part of the IT manager in this situation? First, the manager should continue to monitor the budget all through the fiscal year in order to properly expend the funds allocated in the budget by the end of the fiscal year. A good manager will probably have some "left-over" funds available in the fourth quarter, as these may have been contingencies built into the current year budget. If these contingencies didn't arise during the year, then these funds may be available for other productive reasons. Second, depending on how these extra funds can be used, the manager may be able to improve employee morale by using the funds toward travel, conferences, education and training, hardware/software purchases, and the like. If these funds can legitimately be allocated toward merit bonuses, the employees may enjoy this as well. Last, it is always good to have a cadre of technical people (i.e., contractors, consultants, university professors, etc.) who can be "waiting in the wings" to assist the manager on IT projects (in case extra funds become available). Of course, proper procurement and contracting of these individuals must be followed.

15 Managing Multimedia Development Projects

Will I Be Able To Meet My Next Deliverable and Milestone?

Multimedia development on CD ROMs and over the Web has increased greatly over the past few years. Organizations are using these multimedia programs for education and training, marketing, business presentations, information retrieval, and other applications. I have had the opportunity to act as the project manager on several multimedia projects, where some were produced on CD ROM and some over the Web. This chapter will highlight these projects and discuss some lessons learned.

This has already been done-- let's not reinvent the wheel!

Multimedia Project #1

The first CD ROM project that I managed was geared to helping people learn how to develop expert systems. The CD ROM has been used by over 100 institutions in 31 countries. The CD ROM contained interviews by leading expert system researchers and practitioners, voiceovers, sound, text, graphics, hyperlinks, and animation. The project used Astound as the multimedia authoring package, and this project was the guinea pig for the first multimedia program produced at the college. It took about 1 year of development.

Multimedia Project #2

The second multimedia program was produced on CD ROM and used to help students learn about knowledge acquisition. Knowledge acquisition is the biggest bottleneck in expert systems development, and this program explains some of the issues, techniques, and methodologies to help knowledge engineers learn how to elicit knowledge from the domain expert(s). This program used Authorware Professional 3.0 and took about 7 months to produce.

Multimedia Project #3

The third CD ROM project was a multimedia program that explained the basics of information systems and then used a mystery analogy to allow the student to learn about information systems development methodologies. The mystery analogy allowed the systems analyst to use problem-solving steps in developing an information system in a manner similar to a detective solving the mystery. The program used CBT Express and contained almost every type of multimedia in the program.

Multimedia Project #4

This project resulted in a CD ROM and a Web-based product. This effort focused on developing an information warfare tutorial that would give the user a general introduction to information warfare at a strategic level. The multimedia project team used Authorware Professional 3.0, and took about 1 year of development for the CD ROM and Web-based versions.

Multimedia Project #5

This last project resulted in developing multimedia case studies over the Web. The thrust was to develop case studies via the Web so that the students in the information systems courses could access these cases and interact with them over the Web. The students could also send their case briefs interactively within the case to the instructors. JavaScript was used mostly to build these cases, and the project took about 4 months per case development.

A Few Key Lessons Learned from These Multimedia Projects

From an information technology management viewpoint, there were several lessons learned. The first is that a multimedia project is a team-based approach, synergizing the individual talents of the team members. Typically, there will be a project manager, multimedia authoring specialist, subject matter expert(s), and graphics artist/video specialist. The job of the project manager is to keep this team working in harmony to produce a product that meets the requirements (technical, time, and cost) of the customer. Project management techniques, weekly meetings, communications via e-mail/Web/face-to-face, and interactive reviews of the work help to ensure a successful product.

Typically, the multimedia team is full of artistic, creative thinkers. As the project manager, it is important to channel this creativity and come to closure on multimedia design aspects of the project. By having deadlines and regularly scheduled meetings, this will help force the team to come to agreements to move the project forward.

Another important lesson learned is to make sure that the users and customers are in the loop. As with any information technology project, the users are an integral part of the team. For a multimedia or any IT project, the users provide input during the analysis, design, development, and implementation stages.

Last, the use of iterative, rapid prototyping is a helpful methodology to obtain user comments during the analysis, design, and development process. For multimedia projects, rapid prototyping helps flesh out the user requirements, storyboard, and other design elements.

16 Critical Success Factors in Multimedia Development

Be Careful--The Water is Full of Hungry Sharks, Disguised in Consultant or Lawyer Outfits!

Introduction

Multimedia is just beginning its growth spurt in its use in "edutainment," education, training, marketing, business presentation, information retrieval, and other areas. From an education standpoint, conflicting studies exist regarding the success of multimedia. There have been academic studies showing that multimedia has improved learning retention by over 50%, whereas other studies suggest that the multimedia evaluation studies to date lack rigor and proper experimental control. From casual observation of watching students use multimedia CD ROMs, it appears that they are enthusiastic and enjoy the experience of using the multimedia programs for supplementing their course material. From a faculty's perspective, the multimedia program liberates some of the faculty's time and then the faculty member can increase the sophistication of the course material.

From the author's experience in developing, deploying, and using multimedia programs for educational purposes, there are several critical success factors that have been learned. These will be highlighted in the next section.

Five Critical Success Factors in Multimedia Development

"Know Thy User"

It is critical to know who is the intended user of your multimedia program. If you (as the developer) are assuming a certain amount of prerequisite knowledge for understanding the content in your multimedia program, then this should be considered up front in the planning and design stages of your system. Knowing your user community is important for tailoring the content and presentation of the multimedia material. There may even be different types of users for your system, and this should be taken into account in the design of your system. For example, one multimedia system on "Information Warfare" was developed for three types of users: congressional staffers, division-level military officers, and students. In designing this system, different levels of detail and granularity can be accessed, based upon the user profile.

Incorporate User Feedback throughout the Multimedia Project

As successive prototypes of the system are built iteratively, it is vital to let the users work with these prototypes and then incorporate their comments into the next version of the system. In this manner, the system won't be designed in isolation, and the users will feel actively involved as part of the design team, causing the final system to be something that the users will accept and use.

Users Like Flexibility

Most users of multimedia programs want the system to be flexible enough to navigate at will throughout the system; enable the user to see text, video, and graphics (or hear sound) at the user's pleasure; and contain information at varying levels of detail. The developer should be aware that these requirements come at a cost, in terms of development time and dollars. Some compromise between the developers and users needs to exist in order to design the program in a way that it will be usable. Depending upon the scope, time, and budget of the system, the user requirements will need to be determined and assessed.

Substance and Style Go Hand in Hand

Many multimedia programmers get carried away on the "glitz" of the system and pay less attention to the actual content or message the system should convey. Substance and style go hand in hand, and the developer should make sure there is "something behind" the fancy screens. A well-designed system will include a proper mix of both substance and style.

Don't Be Surprised If You Have to Integrate and Use a Few Packages to Develop Your Multimedia Program

Even if you use a multimedia authoring language, you (as the developer) will probably need to use other software to develop your multimedia program. Software for capturing video, sound, and graphics will be needed to develop the multimedia program. There also may be a need to develop DLLs (dynamic link libraries) in C, Visual Basic, or some other language to augment and enhance the capabilities of the multimedia program. Don't be surprised if "one tool isn't enough!"

Summary

These five critical success factors, as briefly described in this chapter, should be kept in mind when developing the multimedia program. By adhering to these rules, the likelihood of success of your multimedia project will be greatly increased!

17 Lessons Learned in Deploying Multimedia

Think Big, But Within Reason

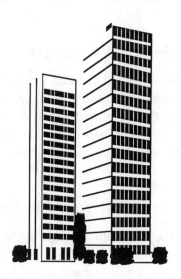

Once a multimedia program is developed, what are some helpful ways of commercializing and marketing it? Five major lessons come to mind from my experiences in distributing multimedia CD ROMs worldwide.

Lesson #1: Know Your Target Market

"Know your audience" is a familiar phrase. No matter whether you are giving a speech or selling a product, knowing your audience or target market is essential.

In one of the CD ROM products that we were distributing, we knew that expert system educators and their students would be the best market. We therefore targeted our brochures for this audience and we used mailing lists of societies that specialized in expert system educators. By narrowing our target market, we saved on unnecessary mailing costs that could have arose from mailing to a very diverse community of users.

After the first mailing of brochures, we found that there was more interest in the CD ROM abroad than in the States. So, we decided to have a second mailing of brochures to the international market of expert system educators to whom we did not send the first brochure. The response rate was tremendous, and we felt comfortable that we found our market niche.

Lesson #2: Provide Some Incentive

We offered quantity discounts and reduced network version prices of our CD ROM to attract further interest and sales in our CD ROM. The educators took advantage of these offers in order for their students to gain easy access to the CD ROMs. The single copy price of the CD ROM was deliberately low in order to be affordable for the educator's market.

Lesson #3: Be Creative in Your Marketing

Instead of just selling the CD ROM through mailing lists, we looked into other ways to further market the CD ROM. One technique was packaging the CD ROM with a book. In this manner, the CD ROM would be a value-added feature to the text book, and the professor and student would enjoy this special deal. The CD ROM was also included in continuing education short courses to further disseminate the CD ROM usage. Besides books and courses, papers and demos describing the CD ROM were presented at conferences and exhibitions to further spread the word about the product. Product reviews of the CD ROM also helped to market the CD ROM.

Lesson #4: Have an Efficient Distribution System

In order to have good customer relations, you should have an efficient distribution system once orders have been received. Delays in receiving the CD ROM once it has been ordered could create customer dissatisfaction. The educators market is typically prone to a semester or quarter teaching schedule. Thus, professors usually need the course materials (i.e., texts, software, etc.) at the beginning of the teaching semester or quarter. Promptness in delivery of the CD ROM, therefore, is essential in meeting the demands of the customers.

Lesson #5: Beware of the Competition

Hopefully, you can develop a product that is unique. Even if it is unique, competitors will soon move in and develop a similar product. You need to be aware of your competitors and their competing products. Look at their prices and their marketing strategies. Compare what you are doing with the competition. By doing so, you hopefully will not be blind-sided by your competition, and your CD ROM will survive!

18 Integrating Multimedia into the Curriculum: Technology Transfer Issues

A Road Map and a Schedule Can Help You Work Your Way Through the Maze!

Introduction

Multimedia is one of the fastest growing technologies worldwide. It has particular relevance in education, training, business, marketing presentations, information retrieval, and the like. One of the areas where multimedia can have a dramatic impact is in the classroom. This has been recognized by a leading undergraduate institution in Virginia, and this vignette will describe how multimedia is being successfully transitioned into the classroom at that university.

The university in this case is James Madison University (JMU), located in Harrisonburg, Virginia. A newly established College of Integrated Science and Technology (CISAT) has been created at JMU to develop collaborative, context-sensitive problem solvers in science and technology. An important underpinning of this new undergraduate curriculum is the incorporation of several key technologies into each of the 4 years of the undergraduate CISAT program. Among these technologies are knowledge-based systems and multimedia. This vignette will focus on the latter technology to show how it is being incorporated into the classroom.

Developing a Multimedia Infrastructure

CISAT recognized the importance of using multimedia in the curriculum to enable the students to have "active" learning, as opposed to passive learning through one-way lectures. In order to develop a multimedia capability, CISAT first established a Center (or Lab) for Multimedia Development. This center was staffed by multimedia specialists who could work with the faculty, administration, and students to develop multimedia programs for the curricula and for college marketing efforts. The center was equipped with the "latest and greatest" equipment and resources for multimedia development. These multimedia specialists conducted workshops and tutorials on how to use the multimedia authoring and presentation packages, and also worked with the faculty to jointly develop multimedia programs for the CISAT curriculum.

The first multimedia program for the CISAT curriculum was a CD ROM called "Developing Your First Expert System." This program was developed in about a year, where the professor in expert systems served as the content specialist and worked collaboratively with one of the multimedia specialists in the center. This multimedia aid serves as an introduction and tutorial to help students and others learn about expert systems development. The program went through several iterations where students at two universities (JMU and George Washington University) used and evaluated earlier versions of the program, over a local area network, in order to suggest improvements for the next version. As a result, this multimedia aid is now being distributed worldwide as a CD ROM through CRC Press. This multimedia aid sparked the interest of other faculty at CISAT, and now there are a number of multimedia projects underway through the help of the Center.

At this point, it is important to note that CISAT provided an excellent way for faculty to become accustomed to multimedia development. By working with the center's multimedia specialists, the faculty could ease into multimedia development by first acting as the content specialist and then picking up the multimedia nuances gradually. The workshops and tutorials offered by the center on the use of these multimedia authoring packages also helped to transfer the multimedia technology to the faculty.

Another successful technique used by CISAT was to purchase multimedia software for each faculty member. The center also provided training on this multimedia software to each interested faculty member. Additionally, each faculty member was given a multimedia-equipped computer and many classrooms were also equipped for multimedia usage.

These techniques were so useful that JMU established a university-wide Center for Integrated Learning Resources wherein one part is a Center for Multimedia. Through this university-wide multimedia center, courses and seminars are given in a number of topics dealing with multimedia.

Lesson Learned

This vignette is an example of incorporating multimedia information technology into a mission and then finding mechanisms to make the mission a reality. CISAT's administration recognized early on that an infrastructure would need to be built within the College if they were to achieve their mission. Many universities do not provide the necessary resources in order to properly institutionalize a technology within the school or university. This was not the case at CISAT. CISAT had the vision to recognize the importance of multimedia in the classroom and then developed ways for facilitating this introduction of a new technology into the college.

19 Legal Issues for Expert Systems Institutionalization

In order to build, market, and distribute an expert system, the organization should be aware of some important legal considerations. For example, is the manufacturer of the expert system liable if the expert system produces faulty advice? Can the manufacturer of the expert system protect his expert system from copying through copyrights and patents? Can the expert whose knowledge is in the expert system receive royalties each time a copy of the expert system is sold? Does the organization using the expert system need a site license to make multiple copies to distribute throughout the organization? Should the expert system be resident on a mainframe and then downloaded to terminals to protect the expert system from tampering? Should the organization use the expert system on a personal computer (PC) and then make copies for the hundreds of users within the organization each time an updated version of the expert system is made?

All of these questions are important to the expert system institutionalization process. Managers must be aware of possible legal entanglements resulting from expert system use, and should anticipate possible legal actions and protect itself against these. Likewise, managers must think in advance of how the expert system is going to be distributed to the users within the organization. Managers must think about the kind of network capabilities that are needed within the firm to access the expert system. Do the managers want to transition the expert system to a separate organization to handle expert system updates and customer support? These types of questions must be answered in the minds of the managers in order to properly institutionalize the expert system. The earlier managers think about these questions, the easier it will be to safeguard against these possible problems.

This section looks at legal issues involving expert systems in order to get management and system developers to think ahead regarding these issues.

Legal Issues Involving Expert Systems

The trend toward mass-marketed expert systems suggests that legal liability will become increasingly important.[1] Expert systems are being designed for the layperson to help draft wills, make contracts, diagnose children's health problems, and provide tax advice. What happens if the expert system incorrectly diagnoses a 5-year-old's medical problem or the tax expert system gives advice that proves faulty after the taxpayer's IRS audit? Who is liable? This results in a tort issue. At one end of the spectrum is total responsibility for the action taker, while at the other end is total responsibility for the producer of the advisory system.[2] One assumption that has been stated by expert system developers is that since someone must act on the results of the system's advice, the person taking action will always have full responsibility. This assumption has not been effective in comparable situations such as navigation charts, books, or medical diagnostic and monitoring equipment.[2] Therefore, it is unlikely that this approach will prove to be an effective defense in the case of expert systems.

It is quite possible that expert systems used in fields such as medicine, law, and accounting may well become subject to licensing requirements.[3] Already, medical expert systems come under the auspices of and are regulated by the Food and Drug Administration. Will an expert system that aids in controlling air traffic have to be licensed by the Federal Aviation Administration, just as a human controller is?[3] This situation may not be too far-fetched. Of course, one sees the critical importance of thorough verification and validation of an expert system!

In the coming years, the next growth areas for litigation relating to AI will be in liability, ownership, licensing, and patent infringement. The issue of liability is a major concern among expert systems developers. Because of the complicated relationship between the vendor and the user, allocating responsibility for such failures is by no means straightforward.[3] And there will almost certainly be failures; no complex computer program has ever been marketed that did not have some defect, somewhere.[3] The issue of ownership, that is, protecting a program developer or producer, is guided by the service vs. product distinction.[4] As Zeide and Liebowitz explain:[4]

> If a program is a product, it is possible that it can be patented. However, this would require that it be deemed a novel invention, unique in the state of the art. It cannot be on the market for more than one year prior to the issuance of the patent. In addition, a patented product is usually a tangible item. But although programs are sold and transferred on physical discs, it is the content of the disc, not the disc itself, that is really the "product." For these reasons, developers usually go the copyright route for protection of their programs. This further substantiates the conflict whether an expert system is a product or a service.

With the possibility for legal suits resulting from AI services and products, how can the AI developer protect himself from possible legal action? This will be explained next.

Ways for the AI Developer to Limit His Liability

First of all, the AI developer should use disclaimers in sales contracts to limit his liability against such things as misuse, or to warn the user regarding the output from the AI product.[5,6,7] For example, an expert system to help diagnose breast cancer might have a disclaimer stating that the conclusions reached by the system are not final and, no matter what the findings, the user (i.e., the patient) should consult her doctor. In the event of incorrect advice, this disclaimer does not get the AI developer "off the hook," but a disclaimer may be used to perhaps limit his liability.

Second, give full disclosure of use of the AI product. For example, a physician using an expert system to diagnose a medical problem should get informed consent from the patient by stating (or having the patient sign a notice in writing) that he is using an expert system.

Third, the AI developer should give notice/warnings to the user. For example, notices should be placed on the building in the robot work area to tell users of the "do's" and "don'ts" when working with or on the robots.

Fourth, the AI developer should do a legal search of copyrights, patents, and trademarks to see what others have done to protect themselves from infringing upon the rights of others and to accordingly protect his product from being copied by other developers/manufacturers.

Fifth, the AI developer should be aware of the use of run-time licenses, with which for a set fee (or certain dollar amount per copy) an AI product can be sold to others by a buyer of that product. For example, if an expert system is being developed using an expert system shell and the user of the shell wants to sell his newly created expert system to others, then he normally needs to pay a run-time license to the expert system shell developer to do so.

Sixth, the AI developer should perform preventive maintenance (as well as corrective maintenance) on appropriate AI products, such as robots. This will help ensure the proper working order of the robot and will limit the risk of accidents to the workers.

Finally, if there is a standard of care for computer use in a particular profession, then a person (i.e., the AI developer) should know clearly what is expected of him and act accordingly. Additionally, the AI developer should purchase professional liability/malpractice insurance appropriate to his field.[5]

References

1. Eliot, L. B. (1988), "Trends: The Commercialization of Expert Systems and Neural Networks," *IEEE Expert,* IEEE, Los Alamitos, CA.
2. Hayes, C. (1988), "The Problems of Artificial Intelligence," *Palm Beach Review,* December 28, 1988, pp. 11-12.
3. Hyman, W. A., W. L. Johnston, and S. Spar (1988), "Knowledge Based and Expert Systems: System Safety and Legal Issues in AI," *Computers & Industrial Engineering,* Pergamon Press, New York, Vol. 15, No. 1–4.
4. Zeide, J. S. and J. Liebowitz (1990), "A Critical Review of Legal Issues in Artificial Intelligence," in *Managing AI and Expert Systems,* D. A. DeSalvo and J. Liebowitz (Eds.), Prentice Hall, Englewood Cliffs, NJ.
5. Liebowitz, J. and J. S. Zeide (1989), "A Little Legal Common Sense for AI Developers," *AI Week,* Birmingham, AL, January 15, 1989.
6. Frank, S. J. (1988), "What AI Practitioners Should Know About the Law: Part One," *AI Magazine,* American Association for Artificial Intelligence, Menlo Park, CA, Vol. 9, No. 1.
7. Frank, S. J. (1988), "What AI Practitioners Should Know About the Law: Part Two," *AI Magazine,* American Association for Artificial Intelligence, Menlo Park, CA, Vol. 9, No. 2.

20 GUESS: A Generically Used Expert Scheduling System for NASA

Jay Liebowitz*
Dept. of Management Science
George Washington University
Washington, D.C. 20052

Chapman Houston, Vijaya Krishnamurthy, Alisa Liebowitz, Janet Zeide
American Minority Engineering Corporation (AMEC)
10422 Armory Avenue, P.O. Box 509
Kensington, Maryland 20895

William Potter
Spacecraft Control Programs Branch
Code 514
NASA Goddard Space Flight Center
Greenbelt, Maryland 20771

Introduction

As some of the previous case studies have been bittersweet in terms of their technical, management, or implementation success, here is a successful story discussing the design of a generic expert scheduling system built for NASA.

* Also, Research Scientist at AMEC.
Acknowledgment: American Minority Engineering Corporation gratefully acknowledges the support from NASA Contract NAS5-38062.

139

Reuse, reuse, and reuse...

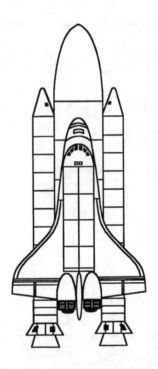

Scheduling is a prevalent function that is omnipresent throughout many industries and applications.[1,2] Whether bus scheduling, job shop scheduling, gate assignment, railway crew scheduling, or other scheduling application, a great need exists for developing scheduling toolkits that can be generically applied to a number of different scheduling problems.[3] Some generic expert scheduling systems are being used, such as Ilog's Schedule/Solver, but more research is needed in pushing the state of the art in generic constraint problem-solvers as related to scheduling.

The U.S. National Aeronautics and Space Administration (NASA) has recognized this need for developing a generic scheduling toolkit. Scheduling is an activity that NASA probably performs 365 days a year. Scheduling is a critical NASA function, whether scheduling NASA shuttle flights, payloads, and crew members or scheduling scientists to use the NASA-supported satellites. Traditionally, each NASA contractor has used different architecture and scheduling techniques for solving their NASA scheduling problems. Since many of these methods can be reused, NASA has been interested in developing a universal or generic expert scheduling system that could be applied to many NASA (and other organizations') scheduling applications. Toward this goal, NASA has developed such scheduling toolkits as PARR,[4] AMP,[5] Plan-It,[6] and others[7] to facilitate reuse of these scheduling architectures and methods.

In keeping harmony with the generic flavor, NASA Goddard Space Flight Center has encouraged the development of GUESS. GUESS has been developed in an object-oriented programming paradigm, with a hierarchical architecture. Its primary purpose is to be used as a generic expert scheduling system architecture and toolkit. The next sections will discuss the design and development of GUESS.

The Architecture of GUESS

GUESS has been designed to take advantage of an object-oriented, hierarchical architecture. GUESS contains two major levels of schedulers. The low-level schedulers are composed of different scheduling methods, mainly heuristic-based and algorithmic-based. The high-level scheduler, called the metascheduler, serves several functions. It first helps in deciding, through an expert systems component, which low-level scheduling approaches are most appropriate for the scheduling application. The metascheduler also coordinates the activation of the low-level schedulers and injects any new information that is pertinent to the scheduling problem.

The low-level schedulers could include heuristic conflict resolution strategies (e.g., shifting, bumping, deleting, etc.), forward and backward loading, optimization routines, neural network algorithms, genetic algorithms, and hybrid techniques (e.g., knowledge-based simulation). For many of NASA Goddard scheduling applications, heuristics are often used by the human schedulers for scheduling satellite experimenters to use the satellites. GUESS is designed to aid the human scheduler and to keep the human scheduler in the loop. As a result, GUESS is a decision support aid, as opposed to an automated replacement for the human scheduler.

In the NASA Goddard environment, scheduling a sequence of activities for a spacecraft is becoming a more complicated process as spacecraft become more complex. The scheduling process must take into account the capabilities of the spacecraft and the trajectory and Space Network information as well as the activity requests from the scientists and from the spacecraft engineers.

Each activity request contains the following information about an activity: the most desired starting time, intervals of acceptable starting times, the duration requirements, and the spacecraft resources that it uses. The resource usage descriptions specify the amount of power required, the data handling required, the instruments needed, and the like. Also included is an explanation of the manner in which the instrument is to be used; for example, the direction in which the camera should be pointing.[8]

Complications could arise from several sources. Resources such as power and data handling should not be oversubscribed. Interactivity dependencies must be met, and forbidden states of the spacecraft must not be entered. The resource requirements that stem from the description of a spacecraft and are needed to support an activity must be included; for example, if the scan platform needs to slew, the power for doing so is figured into the usage display at an appropriate time.

A sequence is considered complete when it is ready for command generation, the process of writing the actual spacecraft commands that achieve the activities in the determined sequence. In order to be considered complete, a sequence must have resolved resource conflicts; that is, the requests to the payload (instruments), power, attitude control, and data subsystems must be feasible. Also, rules involving resources from several systems must be obeyed.[8]

GUESS is programmed in Visual C++ and runs on an IBM PC Windows environment.

THE OBJECT-ORIENTED STRUCTURE OF GUESS

3.1 Implementing GUESS in C++

A schedule has an events list and the resource list. A resource list is made up of resources of different types that directly or indirectly have an impact on the schedule. The events list is made up of different kinds of events. Most of the events have to be scheduled while some act more like markers in the schedule to support the resource modeling.

The primary organizing construct is a class that describes an object. GUESS is a generic scheduling system. It can schedule different kinds of things. The OOPS feature of GUESS is that classes represent various abstractions of scheduling objects, such as events, constraints, resources, etc. An event in the schedule is implemented as a class that has members and functions acting on it. An event has its priority, start time, end time, a constraint list acting on it, an effect's list, and finally has a text-based name for human benefit to improve the readability and maintenance.

All events are inputted with a start and stop time. If an event has time conflicts, the event will be rescheduled. Technically, this qualifies GUESS as a repair-based scheduling system. Always having start and stop times avoids the concept of having to track an event's status as scheduled vs. unscheduled. Also, the many objects in the system that deal with events don't have to know about scheduled/unscheduled status because an event is always considered scheduled.

The constraint list in an event has different kinds of constraints acting on the event. A constraint class has the weight of the constraint and the constraining resource or event. A constraint can indicate its satisfaction level. An event's satisfaction level is computed as a weighted average of its constraints satisfaction. An event's satisfaction is a measure of how well scheduled it is. In addition to after the fact measuring, constraints are also used to generate start and stop times for events.

A constraint by an event is a relational constraint. Relational constraints have the following subclasses: Before, After, During, NotDuring, StartsWith, EndsWith. A relational constraint is time based and can be between any two events. Suppose event E1 has to come before event E2. Event E1 would then have a Before constraint to E2. The other relational constraints operate as expected. Note that the previous case of E1 having a Before constraint does not imply that E2 has an After constraint. If desired, the After constraint would have to be separately created for event E2.

Event classes, constraint classes, and resource classes all have names for user readability and inputting. With inheritance, they can all use the same name-handling code by inheriting from a common parent class. The subclasses of events, constraints, and resources get the same name-handling code by default. Also by inheritance, the list-handling code is used by many classes.

Objects in lists are accessed by the cached reference system. A cached reference to an object has both the name of the object in question and a pointer to the object. The pointer is found once by searching the list. By caching the pointer, the cached reference needs only to look up an object in a list once, saving time. A more important feature of cached references is the ability to refer to an object that hasn't been created yet. Interdependent object systems can have the problem of needing to reference an object before it exists. We don't look up the reference until it is needed, thus allowing a legal reference to an object before it exists.

The Generic Nature of Guess

An object-oriented approach has been used for GUESS in order to maximize the reusability and corresponding generality of GUESS. A library of scheduling test cases is being built with test cases ranging from manufacturing process scheduling, restaurant order scheduling, bus scheduling, to child care scheduling. At the time of this writing, GUESS can schedule, as an example, 2,551 events and over 14,000 constraints in less than 45 seconds on a Dell 486 computer. In the coming months, test results should be achieved in order to determine its generic capabilities. It should be noted, however, that generic scheduling is an ambitious goal, but part of GUESS research is to determine an appropriate way of handling this capability. We hope that the object-oriented, hierarchical structure will greatly facilitate generic scheduling.

Conclusions

The GUESS team is working closely with the end-users of GUESS to ensure that transitioning of GUESS will be a smooth process. There is a possibility of the NIH (Not Invented Here) Syndrome, as GUESS is being built by a contractor, not NASA employees themselves. Hopefully, by working closely with the users, GUESS will be used successfully and not placed on a shelf to collect dust!

References

1. Pinedo, M. (1995), *Scheduling Theory, Algorithms, and Systems*, Prentice Hall, Englewood Cliffs, NJ.
2. Morton, T. and J. Pentico (1993), *Heuristic Scheduling Systems*, John Wiley, New York.
3. Brown, D. and W. Scherer (Eds.) (1995), *Intelligent Scheduling Systems*, Kluwer Publishers, Boston.
4. NASA Goddard Space Flight Center (1994), *Proceedings of the 1994 Goddard Conference on Space Applications of Artificial Intelligence*, Greenbelt, Maryland, May, 1994.
5. Stolte, A. (1994), "An Object-Oriented Approach to Scheduling," *AI Expert*, Miller Freeman Publications, San Francisco, CA.
6. Lee, J. K., M. Fox, and P. Watkins (1993), Special Issue on Scheduling Expert Systems and Their Performances, *Expert Systems With Applications: An International Journal*, Elsevier/Pergamon Press, New York, Vol. 6, No. 3.

7. Zweben, M. and M. Fox (Eds.) (1994), *Intelligent Scheduling*, AAAI/MIT Press, Cambridge, MA.
8. Collins, C., J. Geroge, and E. Zamani (1989), *Strategies for Automatic Planning: A Collection of Ideas*, NASA Jet Propulsion Laboratory, Pasenada, CA, May 1, 1989.

21 An Applications Experience of Introducing EVIDENT to Law Professors

Janet S. Zeide, Esquire
Law Office of Janet S. Zeide
966 Farm Haven Drive
Rockville, Maryland 20852

Jay Liebowitz
Dept. of Management Science
George Washington University
Washington, D.C. 20052

Introduction

Legal expert systems are slowly moving into the mainstream of society. Innovative corporations have developed legal knowledge-based systems in many domains, including commercial real estate and banking law, risk analysis of litigation, and legal and tax issues involving the Internal Revenue Code. The success of some of these systems suggests that legal expert systems (1) save time and legal expense; (2) avoid costly errors that could result in liability and litigation; and (3) contribute to increasing the volume of legal work without a corresponding increase in staff while obtaining an end project that is consistent, accurate, and reliable.

Beware of Those Who Are Afraid of Technology!

One of the most widely known "real world" legal expert systems is Metropolitan Life's CLINT (Check List for INcome loan Transaction). Even though there are successfully employed legal expert systems like CLINT, there is still resistance demonstrated by attorneys in accepting expert systems technology. It appears that attorneys will continually review and analyze extensively the value of developing and using legal expert systems.

Along this point, this chapter describes the applications experience of introducing a legal expert system, called EVIDENT, to law professors at a major university in the Baltimore–Washington, D.C. area. Even though more and more attorneys are accepting the use of computers, is appears that expert systems technology used in the legal domain is too amorphous and may seem threatening to attorneys and law professors. The following will highlight some of the experiences gained from introducing EVIDENT to law professors.

"EVIDENT":
An Expert System Prototype for Determining
Admissibility of Evidence Under the Federal Rules

One of the most difficult law school courses is *evidence*. This area presents hardships for the law student primarily because of the numerous rules of evidence to remember, as well as the need to recall their seemingly countless exceptions. Computer programs, particularly expert systems, could be used as a supplemental learning tool for the student. Some strongly feel that the designers of computer systems to be used in education should take account of the subject of artificial intelligence, and the users of such systems may expect them soon to provide facilities considerably more sophisticated than those available today. Studies have indicated that the computer can be a tremendous training aid.

Based upon these conclusions and the hope of making evidence easier to learn for the law student or young lawyer, an expert system, called EVIDENT, has been developed to determine admissibility of evidence under the federal rules of evidence. EVIDENT runs on an IBM PC compatible.

To test the feasibility of having an expert system for determining admissibility of evidence, an expert system shell, namely Exsys Professional, was used. EVIDENT was constructed using the rapid prototyping, iterative process of knowledge acquisition, knowledge representation, knowledge programming, and knowledge testing and evaluation.

Knowledge acquisition consisted of first having the knowledge engineer read various selected texts and notes on "evidence." A basic understanding of the legal terms and concepts associated with evidence was an essential part of the knowledge engineer's learning process, because it helped in understanding the explanations and discussions of the "expert," as well as aiding the knowledge engineer in asking the right questions when interviewing the expert. For the development of this expert system, a primary expert was used during the knowledge acquisition stage. A high-level overview of the attribute hierarchy of EVIDENT, based upon interviews with the expert, is shown below.

- Admissibility of evidence
  - Relevant
  - Material
  - Legally relevant
    - Hearsay
    - Best evidence
    - Impeachment of a witness
    - Refreshing collection
    - Presumption
    - Writings
    - Opinions
    - Parol evidence
    - Privilege

After acquiring the knowledge through discussions with the expert and using various authoritative texts and bar review notes on evidence, the knowledge was then represented as production rules. The field of evidence lent itself well to situation–action rules, or IF–THEN rules. They are inherently *rules* of evidence.

Through the use of Exsys' knowledge base editor, EVIDENT's knowledge base was encoded. The knowledge engineer found Exsys to be an extremely easy tool to use for encoding the knowledge base.

Upon performing knowledge encoding, the knowledge base was then tested for accuracy and utility. The explanations given by EVIDENT were helpful in identifying problems with the knowledge base and chaining of rules. As expected with expert systems development, problems surfaced such as (1) exceeding the maximum length of a qualifier (30 words), (2) incorrectly writing the rules by including either wrong information, misspellings, or incorrect "and" or "or" expressions, (3) obtaining incorrect output as a result of applying a combination of rules, and (4) having ambiguous (to the user) qualifiers. The iterative process of knowledge acquisition, representation, encoding, and testing was then performed to improve the knowledge base and resulting expert system.

Organizational Adoption Issues Resulting from EVIDENT

EVIDENT was explained and demonstrated to a group of law professors, upon the request of the dean of this major law school in the Baltimore–Washington, D.C. area. The acceptance issues of EVIDENT will be explained next, and can be grouped under the following major headings: fear of the unknown; power struggle; fear of information technology/resistance to change; disbelief in the technology; and replace versus support.

Although lawyers are traditionally viewed as aggressive, and law professors are famously fearless, this group seemed worried. In casual conversation over lunch as well as during the presentation, the law professors repeatedly voiced concerns about being replaced by machines. They became openly hostile in their questioning about why such a system was needed, and how it could possibly help law students or the legal profession.

Fear of the Unknown

There are a number of books that law students typically use as learning aids. Some are "hornbooks" with good reputations, but many, such as *Emanuel on Evidence*, *Nutshell* series, and *Smith's Review*, are not as distinguished but are easier to use and thus more popular. It was explained that EVIDENT would combine the best characteristics of the two existing kinds of study sources, i.e., the program is based on the so-called "hornbooks" but presented in a study-effective manner. This audience was more than reluctant to accept a new method of learning the same "old" information.

Power Struggle

Many of the professors specifically asked — and in various ways — how this program would replace them. They inferred that some loss of their power in the classroom and some loss of control over the students would result. From their reactions, it seemed they were actually worried that they may even eventually be replaced by "machines." They wanted to know what other areas of law were being made the subject of expert systems and then challenged the feasibility of such applications.

In fact, since the law school dean had arranged for the presentation, the professors may have felt there was a hidden agenda. Was the dean trying to "hint" that the school was considering some progressive ideas and teaching methods that might eventually exclude them? Although teaching is normally a rather secure profession, especially for tenured professors, they may have suspected the introduction of a new technology as a means for squeezing them out. One professor even spoke in terms of moving toward the 21st century and the future shock that society is experiencing.

Fear of Information Technology

The law professors attacked the whole concept of encoding the information onto a computer program. They noted that the rules of evidence were most strategically used in the courtroom and that it required a "human touch." They kept referring to EVIDENT as a "machine" and as "technology." They certainly didn't seem to be able to picture a computer program emulating a human expert. The presentation included viewgraphs of several sample sessions with the program. They were anxious to find errors in the program and challenged the material before all of it was demonstrated.

Replace Versus Support

The law professors focused on the system itself rather than accepting that the contents would be provided by a human expert and be taken directly from accepted, existing, published works. They couldn't seem to see that the program would really be another way of helping the student learn the rules of evidence. The law of evidence lends itself so well to expert systems because it is made up of rules and because it primarily requires rote memorization and application. While law professors provide the essential aspect of teaching the ways the rules can and should be used and applied, it is up to the student to memorize the rules and exceptions. EVIDENT provides an interactive study method that would facilitate the memorization process and would serve to reinforce and strengthen the in-classroom discussions and studies. This was explained to the professors, but they refused to view it as a support of their work.

Summary

Many studies still indicate that the top problem preventing widespread use of expert systems is the failure of management to be aware of the technology. Creating an awareness of expert systems is the first step toward organizational development and acceptance of expert systems technology. In the legal domain, as pointed out by this short case study of law professors' reactions, gaining acceptance of a new technology such as expert systems may be especially difficult for legal applications. More research needs to focus on the management and institutionalization of expert systems, including how best to generate interest and awareness of the technology and then ways for properly implementing the technology within the organization. Perhaps, as pointed out in this mini case study, this is the reason why legal expert systems have not been developed in great number and why few legal expert systems are being operationally used in organizations worldwide.

22 Publish or Perish — A Publisher's Dilemma

Often, people aren't ready to accept a new technology!

Prologue

This vignette involves a fairly new international publishing company that specializes in scientific journals, books, and conference proceedings and publications. This publishing company had been involved in publishing the proceedings for an international congress in 1991, and now was embarking on publishing the proceedings for the same international congress scheduled for 1994. There would be about 220 refereed papers from over 45 countries presented at the congress. This amounted to over 1500 pages of text and graphics.

The big question here was to try to be a leader in the publishing field and publish the proceedings in a CD ROM format, instead of hard/soft copy proceedings. As more and more individuals were buying computers equipped for multimedia and CD ROM drives, it seemed that the trend would be eventually to have proceedings published on a CD ROM. The next section describes some of the considerations in determining how the publisher would publish the proceedings for this congress.

CD ROM or Otherwise?

The inclination for the publisher and the organizing committee was to publish the proceedings in CD ROM format, as it would be very unique for an international conference to have such a proceeding in 1994. This would also establish the publisher and the congress as being leaders and on the forefront of this technology.

To best determine the right approach, several factors were considered. Cost was the first factor, and it appeared that the cost for developing the CD ROM master and duplicates would not be much higher than the cost of actually printing hard copies of the proceedings. The second consideration was the necessary format that all the authors had to follow in order to ease the development of the CD ROM. Very specific guidelines were sent to the authors of accepted papers, and they had to follow these guidelines closely. Also, since papers were being sent from all over the world, computer viruses had to be checked by the publisher to ensure that the data on the CD ROM would not be damaged or destroyed. It turned out that there were numerous viruses on the discs sent by the authors, and these viruses had to be removed. A third factor dealt with the availability of the authors worldwide to have access to computers that had CD ROM drives. Over 45 countries were represented at the congress, and many of the authors and attendees from developing nations did not have CD ROM-equipped computers. If the proceedings were only published in CD ROM format, the authors from the developing nations especially would be at a disadvantage. However, printed copies of the proceedings could also be sold by the publisher in addition to the CD ROM copies. The final criterion was portability of the CD ROM for traveling vs. having three to four heavy volumes of the printed proceedings to lug or mail home.

The Decision

The decision made by the publisher and congress organizing committee was to go with the CD ROM format for distribution of the proceedings to the authors and congress attendees. This decision was generally well received by the authors and attendees, although there were some strong reservations by some. Since there wern't any printed abstracts of the papers available at the Congress and the attendees did not have ready access to a CD ROM-equipped computer at their side, the attendees could not read the papers in advance in order to decide which sessions to attend. Of course, the plus side to this situation was that attendance was excellent in the sessions because the attendees did not have the papers in advance; this forced them to attend the sessions to find out about the author's research. A number of individuals, mostly from the developing nations, wanted a printed copy of the proceedings instead of their CD ROM copy because they did not use a computer with a CD ROM drive. However, by and large, this decision to use the newer technology, such as the CD ROM format, was favorably received by most of the attendees.

This vignette shows that there are many considerations involved in introducing a new technology or "bucking the trend." In this case, the publisher and the congress wanted to be a leader instead of a laggard in terms of using CD ROM technology for the proceedings. They were one of the first to use such technology for a conference proceedings, even though there was some resistance to change by the authors.

23 Information/Expert Systems Failures: A Case Study of "Everything That Could Go Wrong, and Did"

Introduction

Information systems (IS) traditionally have played a subservient role to the business functions within an organization. However, times are changing whereby the leading firms have recognized the strategic importance of information systems and have begun to integrate information technology within the strategic vision and business mission of the firm. There are still many organizations whereby the information systems group plays mostly a supportive role, and the IS group is striving to command greater control within the organization. This chapter describes such a real case and covers many factors that cause IS/expert system projects to fail.

In Bill Gates' book, *The Road Ahead,* he stresses the importance of hiring some Microsoft managers who have dealt with failures and failed information systems projects in previous employment. Bill Gates indicates that it is essential to have these types of individuals who have faced adversity before, and have been able to work through these problems and experiences.

Similarly, in the August 1997 issue of *OR/MS (Operations Research/Management Science) Today,* Arnold Barnett (chairman of the OR/MS Today

Committee) states, "Quite apart from the lessons that emerge from analyzing failures, discussing them can remind those in pain that they are not alone. I'm not sure why, but I place great value on this therapeutic function" (p. 4).

According to various studies, like Stephen Flowers' work at the Center for Management Development at the University of Brighton, only 20–30% of IS projects are deemed successful. Flowers indicates the reasons for failure as: fear-based culture, poor reporting structures, overcommitment, political pressures, technology focus, leading edge system, complexity underestimated, technical "fix" sought, poor consultation, changing requirements, weak procurement, development sites split, project timetable slippage, inadequate testing, and poor training.

According to the December 30, 1996 *InfoWeek*, the main problems that cause IS failures are technology fit, project timetable slippage, and poor user-computer interface design.

As shown, there are many reasons why most IS projects fail. In the next few sections, a case will be presented, based upon an actual situation, that invariably led to failure in terms of project inception.

A Case Dealing with Failed IS Project Inception

This case involves an actual non-profit organization with several millions of members. The case centers around the development of an expert system prototype for helping the organization's membership become better informed with respect to housing instrument considerations. After becoming involved in planning a major expert systems conference, the advocate within the organization thought that this technology could greatly help his organization's membership in becoming better informed on housing options so that there will be improved consumer protection and less risk of scams. The advocate contacted a consultant whose expertise lied mainly in the expert systems area. The advocate spoke with the director of the business unit who would ultimately foot the bill to pay for this expert system prototype. The director seemed to be very pleased with the idea, but felt the information systems group needed to become involved with the possible project, especially since this expert system would be put on the organization's Web site (at least for initial testing and evaluation purposes). The IS liaison between this business unit and the information systems group, as well as the IS liaison between the advocate's business unit and the IS group, were consulted and brought into the picture. At the same time, contractual relationships were being worked out between the organization and the domain expert (who was the top person in the country in this field). The expert wanted to be sure that he wouldn't lose his livelihood if this expert system were evidentially built well enough to replace him. Thus, lengthy negotiations ensued between the expert and the organization, even though the domain expert was very keen on going ahead with the project.

As drafts of contractual agreements between the expert, consultant, and the organization were getting close to final form, the information systems group stepped in and started to become paranoid that they would be "holding the bag," in terms of maintenance and further development of the expert system, once the consultant built and installed the first phase of the expert system prototype. The information systems group had never worked with developing expert systems, and they were previously put into a predicament where they had to maintain an expert system for the organization after the consultant "left town." Thus, the IS group was extremely cautious and wanted the consultant to develop a very detailed analysis, design, and implementation plan under the control of the IS group vs. the business unit (who was paying for the work). Keeping in mind that this was mostly a proof-of-concept effort with limited funding attached to it, the consultant and the business unit advocates felt the IS group was placing unnecessary burdens on the consultant (and was trying to stall the effort). Additionally, about 99% of the IS development work was done internally by the IS group, so outsourcing this work to someone external to the organization made the IS group uneasy.

As the business and other business units had experienced hardships in the past in dealing with the IS group, several alternatives were proposed to lessen the involvement of the IS group in this project. One alternative was to develop the system as a standalone for the PC without putting it on the organization's Web site. Another alternative was for the business unit to buy their own Web server, and put the expert system on their server (vs. the main one controlled by the IS group); however, the Web administration would still need to be performed by the IS group.

It had been almost a year since the project was initially conceived to the point where other alternatives were being proposed to circumvent the IS group. As a result of these delays and little hope in sight, the consultant did not pursue this project any further and the project is still in limbo.

What Went Wrong with This IS Project?

In analyzing this case, there are several factors that contributed to the lack of project inception for this expert system. First, the "NIH (Not Invented Here) Syndrome" played a major role in this project's failure to get started. The IS group was used to developing their own systems, and not having to rely on outsiders. Even though the IS group was going to be part of the development team (at least as casual observers during the knowledge acquisition sessions), the IS group wanted to control the project entirely.

Second, the first advocate who really conceived of this project was not in the business unit that ultimately would pay for this work. Thus, the advocate perhaps didn't have the clout and authority to command that this effort would be done.

Third, the legal department can greatly delay a project. Contractual delays, caused chiefly by the legal counsel's concerns in protecting the organization, can lead to unexpected slippages.

Last, perhaps the technology itself (i.e., expert systems) was getting too much focus, as opposed to having a way to solve the business problem. Many people in the organization didn't understand expert systems and thus felt uneasy.

What Can Be Learned from This Case?

An important lesson learned from this case is that probably most projects fail due to the management of the technology vs. the technology itself. That is, the organizational and political factions can kill a project, even though the project may be a great and worthwhile idea.

Second, it's important to work with the IS group within the organization, but there should not be dual subordination. The business unit that is sponsoring the work should have the final say and should be the key boss.

Third, don't push a technology onto an organization. Develop appropriate ways of solving their business problems, no matter what technology is used. Don't force-fit a technology onto requirements.

Last, some projects aren't meant to be, even though the idea is valid and can generate key benefits. The case described here is one of many examples of this point.

A CODE OF ETHICS

24 Ethics Should Prevail

Have Integrity!

In today's society, with the constant pressures and struggles to survive, many organizations may be faced with jeopardizing their morals and ethics. I will briefly discuss two organizations — a university and a company — that tried to place me in a precarious position.

The University

In order to be a *U.S. News and World Report* top 50 educational institution, a major university felt justified in their ethics to "play the numbers" game that many other top 50 U.S. universities were supposedly doing. Specifically, the university being described was just below the top 50 universities. In order to break this barrier to be in the top 50, it was felt by the administration and many faculty that the GMAT score reported to *U.S. News and World Report* needed to be higher. In order to best meet this goal, it was decided that a subgroup of the entering full-time students' scores (i.e., essentially, the "better" group) would be those reported. The administration and faculty tried to justify this action by reasoning that other schools are doing this, and that eventually they would accept only those full-time students with GMAT scores over the 600 range, so they were moving toward their ultimate goal.

I felt that reporting the "elite" scores was unethical, improper, and unwarranted. I voiced my concern at the department and school faculty meetings, and there were few echoes of support of my view. This silence worried me greatly, and called into question the direction in which our faculty and school leadership were heading.

What can be learned from this experience? First of all, if you don't believe that the organization is following your principles and ethics, then you should probably move to another organization. Second, you could be a whistleblower on those institutions that are following this deceptive reporting practice, but it may take a while to make any inroads. Last, you can try to convince the faculty and administration of the potential improper actions and hope to change their minds; however, this is likely a difficult task.

A Company

Another situation involving ethics surfaced around the same time as this university ethics question arose. A company was interested in getting a multimillion dollar expert system project through the government. Since the company didn't have any real expertise in expert systems, they asked me to be a consultant on the project. I agreed to explore this possibility, and as such, we had a meeting with the government representative and others to discuss this expert system project. During the meeting, I noted that the company said that I would head up this project, if the company were selected for this effort. This was a surprise to me, because as a consultant and due to university regulations, I would only be allowed to work a day a week on this project. In my mind, perhaps they felt the project just needed a day a week of my time to lead the effort (but I thought they were underestimating the level of effort!).

A few days after the meeting, I was told by a friend who would be involved with this work that the company told the government official that I would be working full-time on this project. This was all done without my knowledge and, of course, was untruthful and incorrect. Since I never heard from the company, I told my friend to make it clear to the government official and the company that I would never work full-time on this project (i.e., requiring that I would have to leave my university teaching position). And, since the company was lying, I would never work with this company on this project or any other effort (due to a lack of ethics again!).

The epilogue of this case is still pending. The government official wanted me to work on this project, and thus was going to drop the company from its consideration in working in this effort. Another company was being strongly considered to lead this effort, if funded, and I could serve as a consultant to this company if I wanted.

Lessons Learned

I was quite astonished and disappointed to see the lack of ethics as described in these university and company scenarios. It was even more frustrating to learn that other organizations allegedly possess a similar lack of ethics. How could one respect oneself or the organization where employed if this goes on? It is extremely troubling to personally witness these poor ethics.

The main lesson that I learned is that one cannot respect an individual or organization if lying, deceit, and poor ethics occur. If one cannot change the values of the organization, then perhaps it's time to leave in order to (hopefully) find another organization that shares one's value system.

25 The Consultant's Dilemma: Risk Versus Return

Money Hungry?

The Consultant's Tug of War: Money vs. Feasibility?

This is a true story of a decision involving a high-risk, high-return tradeoff and the dilemma of whether the consultant should accept this tradeoff by becoming associated with the project.

A potential client called a consultant who had expertise in the expert systems field. The consultant visited the prospective client, which was supposedly engaged in a year-long research effort involving the development of a prototype expert system for engineering design. In speaking with the management and technical members of this project, the consultant quickly learned that almost nothing had been done in 10 months, and now the organization was in a crash mode in order to deliver what was promised in just 2 more months. The organization had promised as part of its research proposal (which was granted and funded) to develop an expert system prototype on a workstation that would be integrated with databases and various optimization codes for designing an engineering structure (the application was actually more specific, but for confidentiality reasons, the term "engineering structure" will be used). The organization really didn't have the expert system expertise on the project team, and managers suddenly realized that the expert system prototype would be due in 2 months and very little work had begun on the project. They were willing to pay the consultant a substantial amount of money in order to "save face" and complete the project as promised.

As the consultant spoke with the project members, it quickly became apparent that the project team members really didn't understand the complexity of the tasks involved, especially to complete the 12 month effort in only 2 months! At first, the project team talked about delivering the prototype on a Macintosh platform, and then said it should be delivered on a Silicon Graphics workstation. The project team wasn't clear on how many optimization codes had to be integrated with the system, and how the integration between the expert system and these other subsystems would be handled. Additionally, the project team members first said that they would do the integration work, but then realized that they weren't sure how to accomplish this effort. At first, the project team seemed to have narrowed down the scope of the project by talking about developing a simple rule-based expert system with links to one optimization code and a database. After the second meeting, the project team had increased the scope of the project by changing the hardware platform, linking with three optimization codes instead of one, and linking with a database that was still under development. Also, during this meeting, the consultant sensed that there were some political motivations associated with the project, and it didn't seem that everyone was speaking from the same page.

By the third meeting, the consultant still hadn't committed to this effort, as he was still doing some preliminary investigation regarding this project. During the third meeting, it was extremely evident that somehow the scope of the project kept growing and the timeline for the deliverables was getting shorter. The consultant had some uneasiness and was uncomfortable as to whether the project could deliver what was promised within the designated time frame.

The Consultant's Dilemma

In order to complete the project, the management of this project was willing to pay a high sum of money to the consultant. This amount of money was certainly tempting for the consultant, but the reality of the situation quickly set in. The consultant felt that the project, even with the consultant's help, would not be able to deliver what was promised. Part of the reason for this was that no work had been done during the first 10 months of the 12-month contract. Even though the money was hard to turn down, the consultant felt it was more important to uphold his dignity and his reputation. Also, there would have been tremendous pressure to "crank out" the work within the next 2 months, and it was extremely doubtful that the work could be completed as promised. The consultant decided that most likely his reputation would be tarnished if he were to be involved in this effort, because the project deliverables could not be met. The consultant felt that it wasn't worth taking that risk, no matter how much money was involved. Some consultants may feel differently, but this consultant perhaps had higher standards and integrity than others.

The moral of this story is to be careful of the dangling carrot. Hopefully, most people will agree that integrity and reputation are more important than money!

26

What Clients Need to Know about the Expert Systems Being Built for Them

Don't Fool the Client!

Which Hat Should I Wear Today?

Introduction

Over the years, overexpectations of expert systems technology by the client have developed. This is partly due to companies touting expert systems technology, and even saying that products are expert systems where, in fact, they are really not. Companies have jumped on the hype of expert systems, and in the process, they may have misinfluenced customers along the way. Customers, however, are now becoming smarter and are educating themselves on expert systems by attending conferences, seminars, and short courses, and by reading various AI literature. As Newquist says,[1] "Customers now have more say in what's going on in the AI industry — customers realize they actually have a right to get what they are paying for!"

To help keep expert systems technology in proper perspective for the client, the expert system developer has certain professional social responsibilities to follow. This chapter will describe what the client should expect from the expert system developer.

Client Expectations from the Expert System Developer

Customers of expert systems should be aware of the advantages and limitations of expert systems technology. In part, this should be taught to the client by the expert system developer; the client should also learn about the technology through seminars, courses, conferences, and literature.[2,3,4,5] Clients should be aware of the following professional social responsibilities that should be performed by the expert system developer.

Clients Should Be Aware of the Capabilities of Expert Systems, and Should Not Have Overexpectations

The expert system developer should be very specific and honest in explaining how expert systems might help the client in a particular situation. The expert system developer should not create false hopes for the client, and then deliver something to the client that did not fill the stated promises. The expert system developer should determine if expert systems technology is appropriate for the problem at hand and then narrow down the scope of the problem for expert systems development. An acceptable error rate should be determined by the client for expert system proof-of-concept feasibility. The expert system developer also has a responsibility of educating the client on expert systems, and the expert system developer should not inflate the capabilities of expert systems. The limitations of expert systems, such as the inability to learn and automatically update its knowledge base, should be equally stressed along with the advantages of using expert systems.

The Client Should Make Sure that the Expert System Developer Researches What Has Already Been Developed for the Client's Type of Application by Other People — the Expert System Developer Should Not Reinvent the Wheel on the Client's Money

One of the important steps in developing any application is to survey and analyze what has already been done in that area by others. This step also applies to expert systems. The expert system developer should review and understand what similar work has been accomplished already in the related tasks, and then the expert system developer can learn from the successes and failures of others. The expert system developer should not use the client's money for developing something that previous projects have shown will fail. The expert system developer has an obligation to use the client's money wisely and prudently. In doing this, the expert system developer must perform the necessary step of reviewing the works of others in similar tasks to the problem at hand.

The Client Should Not Be Fooled By an Elaborate User Interface; That Is, the User Interface Should Not Mask an Incomplete, and Perhaps Inconsistent, Knowledge Base

In the expert systems arena, the knowledge principle applies that states, "The power of an expert system lies in its knowledge." The knowledge base, which is the set of facts and rules of thumb about a particular task, is where the power of an expert system is derived. If the knowledge base is incomplete, then the advice from the expert system will be incomplete, no matter how efficient the expert system's control structure (i.e., inference engine) works.

The user interface (i.e., dialog structure) of an expert system is also very important in the ultimate success, implementation, and use of the expert system by the client. However, the expert system developer has an obligation to not fool the client into thinking that the expert system is very powerful by only developing a sophisticated user interface. In other words, the expert system developer should sufficiently develop the knowledge base, and not hide an incomplete knowledge base behind the mask of an elaborate user interface. The expert system developer should be up front with the client and tell the client what knowledge is actually "hidden" behind the user interface.

The Client Should Know that an Expert System Will Not Always Produce Correct Results

An expert system will not be correct 100% of the time, in the same manner that a human expert is not 100% correct all the time. The expert system may reduce the likelihood of errors being made and it may standardize the approach of solving a particular problem. But the client should not be fooled into thinking that the expert system will give correct results all the time. This needs to be told by the expert system developer to the client.

The Client Should Make Sure That the Expert System Is Properly Verified, Validated, and Evaluated

The expert system should be verified, validated, and evaluated. The expert system developer should check for consistency and completeness of the knowledge base. The expert, other experts, and users should be involved in the validation and evaluation stages. A representative set of test cases should be used to validate the expert system. The expert system developer should check to see if the expert system produces results within the acceptance rate determined by the client. Typical users of the system should be involved in the evaluation stage of the expert system development process. They can help the expert system developer determine if the system is easy to use and easy to update, and if better explanations and descriptions of terms and conclusions used in the expert system are needed.

The Client Should Know That People Will Be Needed to Maintain the Expert System After It Is Built

Once an expert system is developed for the client, there will need to be someone or a group of individuals to maintain the expert system. The knowledge base will need to be updated, the user interface may need to be enhanced, and the control structure may need to be adjusted. Some of the PC-based expert systems may require only one person, half time, to maintain the expert system, while other larger systems may require several individuals full time to maintain the system. For example, XCON, an expert system for configuring VAX computer systems, had about 15 people maintaining it full time. The client should know that there will be maintenance costs associated with the expert system even after it is built by the expert system developer.

The Client Should Expect Proper Training and Documentation from the Expert System Developer on Using the Expert System

An important part of the delivered expert system is the training and documentation on using the system. The expert system developer should tell the client up front that training will be an integral part of the "delivered package." The expert system developer has an obligation to provide training so that implementation and use of the expert system will be achieved. Also, the expert system developer has a social responsibility of providing good documentation on how to work the expert system. Of course, costs for training and documentation need to be included in the contract by the client.

The Client Should Know that Expert Systems Are Not the Cure-All to End All

Many expert system developers have a mind-set that expert systems technology is the only way to approach and solve a problem, even if the client thinks that expert systems should not be used. The expert system developer has a social responsibility for exploring all possible ways for solving a client's particular problem, of which expert systems may be one alternative to consider. The expert system developer must assess whether a problem is applicable for expert systems development, and he/she must also assess if there are better ways of solving the problem. Perhaps, conventional database management systems technology is more appropriate for solving a problem than using expert systems. The expert system developer has a moral obligation to consider many possible approaches to solving the problem, not just concentrating on expert systems.

The Client Should Expect Deliverables to be on Time, at Cost, and within Technical Specifications

The expert system developer has a moral obligation to properly estimate costs, time, and technical specifications for an expert systems development. The expert system developer should adhere to these parameters to the best of his/her ability.

Conclusions

Integrity is an important trait that professionals should possess. With a new technology, the public is less informed of the technology's shortcomings than those of an established technology. Since expert systems is a relatively new technology, the burden initially should be placed on the expert system developers to educate their clients on the advantages and limitations of the technology. This education process is part of the professional social responsibilities of the expert system developer. To ensure the "faith" and success of the expert systems field, expert system developers need to be cognizant of their social responsibilities to their clients and the public. If the expert system developers adhere to these professional ethics, then the client will be better aware of the usefulness of expert systems.

References

1. Newquist, H. P. (1987), "Circling the Wagons: The State of the AI Pioneers," *AI Expert,* Miller Freeman Publications, San Francisco, CA.
2. Liebowitz, J. (1988), *Introduction to Expert Systems,* McGraw Hill, New York.
3. Liebowitz, J. (Ed.) (1988), *Expert System Applications to Telecommunications,* John Wiley, New York.
4. Liebowitz, J. (1987), "Common Fallacies About Expert Systems," *Computers & Society,* ACM, Vol. 16, No. 4 & Vol. 17, No. 1.
5. Cantu, F., J. Liebowitz, and R. Soto (1998), Proceedings of the 4th World Congress on Expert Systems, Cognizant Communication Corp., Elmsford, NY.

FUTURE CONSIDERATIONS

27 Intelligent-Aided Multimedia: Prospects and Issues

Intelligent Systems With Multimedia Makes Sense!

Introduction

Intelligent systems and multimedia are two fields that are rapidly expanding in terms of commercial application and usage. These two technologies are complementary, and could lead to great synergy. Intelligent-aided multimedia is an emerging area that takes advantage of coupling intelligent systems technology to multimedia. Multimedia may be supported by expert systems (i.e., making multimedia more intelligent) in the following ways:[1]

- Supporting a mix of media tools/outputs;
- Designing a mix of text and graphics;
- Using an expert system as a guide for hypermedia;
- Using an expert system as a guide for hypertext;
- Using expert support to imaging systems.

Expert systems may support a mix of media tools/outputs by using the expert system to assist users in selecting, retrieving, and manipulating multimedia information. The system reasons about user input and chooses intelligently among media alternatives. In the second application, an expert system may be used to develop a customized mix of text and graphics for the intended audience and situation. Guidance would be provided by an embedded knowledge base. In the third application, an expert system may provide systematic assistance to help users navigate through accumulated multimedia knowledge. The expert system would provide users with options and evaluate inputs. In the fourth application, expert systems can assist multimedia by leading and directing the user as a guide for hypertext. Expert systems enable powerful knowledge representation and control that would be helpful for multimedia enhancement. In the last application, an expert system could provide control and management support to image and documentation technology.[1]

In order to further develop the synergy between expert systems and multimedia, there are a number of research issues that still need to be addressed. The following sections will highlight some of these issues.

Research Issues

Design/Development

There are various research issues associated with the design and development of intelligent-aided multimedia. Some of the more important issues will be highlighted next.

Developing with vendors more robust integration tools for application design, development, and use

Currently, many multimedia authoring languages have weak links to external programs, typically through DLLs (dynamic link libraries). Often, it is awkward and sometimes difficult to link these multimedia authoring programs with intelligent systems. For example, at the U.S. Army War College, an intelligent multimedia application called MENTOR uses Authorware Professional 3.0 and a CLIPS DLL in order to integrate an expert systems component with the multimedia program. Unfortunately, developing and integrating the CLIPS DLL with Authorware was fairly arduous because of the lack of a robust and easily facilitated structure for accomplishing this integration. In the future, more multimedia vendors should develop robust integration tools to help integrate intelligent systems with multimedia.

Investigating best modes of integration

There are several different software architecture models for handling integration between multimedia and intelligent systems. These are stand alone, translational, loose coupling, tight coupling, and full integration.[2] Stand alone refers to independent modules, independent construction, and no interaction. Translational refers to independent development, delivery and development emphasis, and beginning with either system (i.e., expert system or multimedia program). Loose coupling uses inter-module communication via data. Tight coupling refers to application decomposition into separate independent modules with inter-modular communication via parameter or data passing. Full integration means that the expert system and multimedia program share data and knowledge representation, and there is communication via a dual nature of structures via cooperative reasoning.[1] More research is needed to determine the sets of conditions under which each software architecture is most appropriate, and to determine the ways in which we could move toward the full integration model.

Investigating the feasibility of using interactive multimedia to intelligently support groupwork, group DSS, and management support systems

Collaborative problem solving and groupware, such as GroupSystems 5, are useful for helping to organize, structure, collate, and integrate ideas for decision making. However, many of these systems don't have either multimedia or intelligent systems capabilities. Intelligent-aided multimedia could enhance groupware by having intelligent agents working in the background to help guide the user in making suggestions to improve the decision-making process. The multimedia aspect may help provide some additional explanatory information via videos, graphics, animation, and sound.

Implementation

Besides design and development issues, implementation issues are also critical for the acceptance and success of a new technology such as intelligent-aided multimedia. Some of the major implementation issues are addressed next.

Determining how best to use existing and future information networks and client servers for multimedia applications

As the Web increases in usage and popularity worldwide, many multimedia developers will opt for putting their multimedia materials on the Web vs. distributing them on a CD ROM. For now, the Web is still slow for viewing and downloading video clips, but this will change soon as telecommunications technology advances. Additionally, we are already seeing expert system vendors developing Web-based server versions of their expert system shells, like Exsys Multilogic's NetRunner. In the years ahead, we will need to address the best implementation strategies for delivering these intelligent-aided multimedia programs.

Investigating the impact of intelligent multimedia systems on people and organizations

More research is needed in assessing, evaluating, and validating the impact of intelligent-aided multimedia systems in terms of user and organizational acceptance. Evaluating the effectiveness of multimedia has been a popular topic in recent years, and structured methodologies need to be developed for validating and evaluating intelligent-aided multimedia. It is still not clear if intelligent-aided multimedia programs have a higher acceptance by users than regular expert systems. More studies are needed to look at these issues.

Developing an appropriate cost–benefit framework for the justification of such systems

In order for companies to invest in the development and usage of intelligent-aided multimedia, these software programs must bring value-added to the organization and to the bottom line. Because intelligent-aided multimedia is relatively new, a cost–benefit framework needs to be established for justifying the use and development of these systems. Intangible benefits are often difficult to quantify, yet there needs to be a way of representing these values. In the years ahead, implementation and adoption of these systems will be heavily dependent upon the cost–benefit frameworks established.

Integrating Neural Networks with Multimedia

Neural networks may be coupled with multimedia to identify user patterns and suggest better ways of navigating through the multimedia program according to the user profile and user actions. In this regard, possible input variables could include the following:

- User Profile-Related Variables: the organizational affiliation/background of the user, user's general interest, depth of user's knowledge, specific areas of interest expressed by the user, user's familiarity of hypermedia environments, etc.
- Hypermedia-Related Variables: number of revisited sites, number of hypertext links hit, length of time during each multimedia module, number of correct and incorrect answers/responses according to the module post-tests, length of time spent at each hypertext link, number of sequential jumps from hypertext screen to screen, etc.

The output from this embedded neural network would indicate to the user which multimedia modules are best to peruse, in which order, and how detailed to go into each module.

In the case of the Information Warfare (IW) multimedia aid, developed at the U.S. Army War College,[3] a neural network was being considered for matching various patterns against how well the user answered questions dealing with military-related, technological, organizational, political, social, and historical IW issues. The neural network would take the user profile information and the hypermedia-related information and adaptively suggest where the user might want to proceed in the multimedia program, according to the user's responses, interests, and hyperlinks. In this regard, the neural network would be set up as a DLL to the Authorware Professional 3.0 multimedia program.

References

1. Ragusa, J. (1996), "Tutorial Notes on Integrating Expert Systems and Multimedia," *The Third World Congress on Expert Systems,* International Society for Intelligent Systems, Seoul, February 5-9, 1996.
2. Liebowitz, J. (1993), "Roll Out the Hybrids," *BYTE,* McGraw Hill, New Hampshire.
3. Hluck, G., J. Liebowitz, and R. Minehart (1996), Information Warfare Multimedia Aid, U.S. Army War College, Center for Strategic Leadership/Knowledge Engineering Group, Carlisle, PA.

28 A Snapshot of The Robotics Age

The Robotics Age is Next!

As we move beyond the Knowledge Age, the Robotics Age will descend upon us around 2010. This age will find the permeation of robotic usage at home, in the workplace, at play, and throughout society. This chapter will highlight important areas where the Robotics Age will endure.

Modern Life at Home in the Robotics Age

The year is 2010, and automation has finally permeated the home environment after a long expectation period. It's not quite *The Jetsons,* but the "Robotics Age" has certainly begun to make an impact on today's society.

"I'm home," yells Jason. "Where is Mazel?" (the Springer Spaniel who is usually there to greet Jason).

"Rosie is walking Mazel out front!" replies Mom. Rosie is the newest of the line of domestic robots that can walk your dog in a fixed location, sweep the house, serve drinks, and the like (as long as you can program, using the speech-understanding user interface, the coordinates of objects in her path). Rosie uses computer vision, but even in this year of 2010, requires additional programming for sensing such things as dynamic objects.

"Is Kenny home yet?" asks Jason, looking to play ping pong after school. Rosie plays, but often misses the shots because it takes too long for her to react to the moving ball. Rosie, however, is pretty good at slam shots if you give her enough time through some high lobs.

"No, Kenny is at a friend's house today," Mom shouts from her study as she receives a fax from her sister. Today, every house of Mom's and Dad's family has at least one fax machine, so all the brothers, sisters, parents, and grandparents fax as a daily part of life.

Jason, after not having Kenny to play with, decides to eat a few chocolate chip cookies and play with Mazel as Rosie enters the house with Mazel on the leash.

"Hi, Rosie," Jason remarks while eating a chocolate chip cookie.

"Hi, Jason — how was your day at school?" Rosie replies in a staccato manner through her speech recognition and understanding controls.

"I had a great day," Jason tells Rosie as Mom enters the kitchen.

"Hi, handsome — sorry I was late coming into the kitchen. I was expecting an important fax that I had to reply to immediately," says Mom.

"Hi, Mom," petting Mazel at the same time. "I had a great day at school, acing my math and spelling tests, and beating my all-time record on Computer Minute!" joyfully exclaims Jason.

"What is Computer Minute, again?" asks Mom.

"Computer Minute is where you and the computer each try to get 30 math questions correct, but the computer only gets 10 seconds and we get 60 seconds. Usually, the computer wins, but this time I won because I scored a perfect 30, and the computer scored 29 (only because the computer ran out of time)," Jason says smiling.

"Hi-ya (no that's not saying hello with a Southern accent; it's Kenny doing a karate move as he yells, entering the house)! I'm home," shouts Kenny.

"Where is my karate belt?" Kenny asks, as he heads for the chocolate chip cookies.

"You better ask Rosie, because she did some cleaning today and I'm not sure where she might have put your belt," replies Mom.

"Rosie, do you know where my karate belt is?" asks Kenny.

"Sure, it's right here." Rosie lifts her right arm, and extends it 4 feet to grab the belt from the top shelf in the family room's chest. These extendible, flexible, and durable robotic arms are great!

After Kenny does a few more karate moves, he and Jason go into the family room to relax a bit before practicing the piano, doing homework, and eating dinner.

The 2010 entertainment center is a wondrous thing! It is an 8 foot by 8 foot computer–TV–phone–fax–videoconferencing–entertainment screen tied into the "all-purpose" outlet so that it could be used for such things as a computer for tying into the Web through wireless communications to watching up to 500 channels of television to conducting a videoconference or video-telephone call with your family, friends, or colleagues.

Jason and Kenny decide to use the scanner, which has been preprogrammed to Jason and Kenny's tastes, for selecting the most enjoyable television channel/show to watch. Jason and Kenny decide to override the preprogrammed choices and enter the words "comedy," "cartoons," "adventure," and "Rocky Robots" (the newest character craze for children), and they then press "Enter." The scanner immediately uses these keywords and locates the television show that is currently playing that matches most of these keywords.

Jason and Kenny think that this certainly beats searching through 500 TV channels!

Of course, Jason and Kenny want to watch different shows, so they use the windows effect which breaks the screen into halves, each playing what Jason and Kenny want to watch, respectively. They also wear headsets so they can tune into only their show, without disturbing the other person. And, of course, to add to their comfort, the computer-controlled house temperature settings automatically adjust to offset the outside temperature for ease and comfort.

"What a life," Jason remarks to himself.

The Military in the Robotics Age

By 2010, robotics have advanced to the stage where they are being used in a variety of applications, especially in the military. Picture the following scenario.

In the United States, the Army has diminished in size due to fewer super-power military threats, budgetary cutbacks, movement toward peacekeeping operations, and stronger coalition forces. Even though the number of military personnel has decreased since 1996, there has been an increase in the number of military "robotic" personnel. Robots are being used in the military in a variety of ways. Robotic "pilot" clones to fly decoys, robotic tanks, robotic firefighters, and robotic servicepersons to try to detect (and possibly trip) land mines are examples of where robots are being used in the armed forces in the year 2010.

The use of robots is ideal in situations that are perhaps too dangerous for humans. Mobile robots could be used as sentries in the military. Robotic arms are being utilized to help in situations where there might be fires aboard ships or to help manipulate tank controls.

Aside from these fairly large robots, an enormous growth has been in "microrobots." These are low-cost, small, disposable robots, that can be thrown away or left behind when they finish their task or run out of power. People are comparing them to ballpoint pens: cheap enough to toss out when they break. The microrobots are one-chip robots with sensors, computers, actuators, and power supplies on board. Microrobots are being used to fix things such as a break in an underground electrical conduit or, through silicon micromachinery technology, to create micromechanical motors for possible medical applications. The thrust in the year 2010 is using an entire robot system on a chip, which allows mass production using integrated circuit fabrication technology and also allows costs of production to greatly decline.

Even though it might be hard to think of a robot as a comrade and friend, it certainly is preferable to have a mobile robot moving through a field of potential land mines than it is for humans to risk their lives doing the same task. In this regard, the robot has become a valuable addition to the armed forces personnel!

Business in the Robotics Age

There seem to be new high-tech firms sprouting everywhere in 2010. Part of the reason for this growth is due to the robotics manufacturing and application fields. Speech understanding and computer vision have improved to the point where robots can understand spoken commands with a fairly comprehensive vocabulary. Through advancements in telecommunications, real-time control (even in a distributed environment) for robotics locomotion and operation has vastly improved from the previous decade.

Robotics in the workplace have become more affordable and cost-effective. Automobile manufacturing plants continue to use industrial robots for their manufacturing processes. Ship building and repair firms use robots for assembly, painting, welding, and other tasks amenable for industrial robots. Even smaller manufacturers, such as appliance manufacturers and tooling companies, are using robots to help them in their manufacturing processes.

At the office, robots are in active use in a variety of ways. Robots are being used (within union laws) for cleaning and sweeping the office floors and corridors. Robots, through the help of intelligent agent technology, are also being used for providing routine office functions (like building security monitoring for possible break-ins and intruders). Pick-and-place robots are being utilized at the loading docks for loading and unloading operations at buildings and warehouses.

The nice thing about robots is that they don't complain, they work hard, don't take coffee breaks, don't take vacations or annual leave, don't strike, and haven't unionized. Labor unions, however, still carefully monitor the percentage of robotics used in the workplace so that human jobs are not lost; but if this does occur, labor unions will ensure that those individuals should be retrained, retooled, or re-educated.

Internationalization in the Robotics Age

Robots in 2010 are multilingual. Even though the cost of a robot has decreased over the years, they are being used primarily in developed, industrialized nations. The developing countries have a surplus of cheap labor and can't afford the use of robots in the workplace. The Pacific Rim nations (e.g., Japan, Korea, etc.), Europe (most notably Germany and France), and the United States are the primary buyers and developers of robots.

An untapped market is South America and Eastern Europe. In South America, robots could be used to help mine the precious gems and also help in agriculture and harvesting the fields. In Eastern Europe, industrialization is surging where new construction and foreign investment have proven to help rebuild the Eastern European block countries. Robots could be used here to help with building construction and maintenance.

Robots, with their multilingual capability, are also being used as interpreters to help with foreign business translation and negotiation. Robots are still being used heavily in manufacturing, and their growth is expanding worldwide.

Medicine in the Robotics Age

Medicine in the year 2010 has advanced remarkably over the past decade. Due to genetic and biomedical engineering research, the life expectancy of an individual has increased to the late 80's, from the 70's just a decade ago. Part of the reason for longer lives is the use of robotics in medicine.

Biomedical engineering is using robotic limbs, micromechanization of robots in surgeries, and even robotic vision components for improved eyesight. The idea of the "bionic" person is still a distance away, but we have moved dramatically closer to this realization. In the operating room, robotic aides to assist the doctor in routine tasks are being used.

In typical medical offices, doctors haven't been using robots extensively, possibly due to malpractice issues and the "human element" being eroded through the use of robotics and automation. However, in spite of these concerns, some robotic arms are being used to give shots, instead of involving the nurse or doctor in this routine activity.

In the future, microrobots will be used increasingly in surgical operations. Material science will be an important contributor to developing robotic components used as part of the human body.

Sports and Recreation in the Robotics Age

Robots will not have advanced to the point in 2010 where they are generally accepted by sports lovers to be team members and participate in sports. However, robots may be well accepted by athletes and spectators in terms of their training routines. For example, a robot to fire tennis balls and possibly return shots is very possible by 2010. Additionally, robots for playing ping pong, possibly serving as goalies, and used for kicking balls may be commonplace in the practice regimen of athletes in 2010.

For recreational purposes, robots may be used in chess, checkers, cards, and other logic-driven games. They could also be used in pitching and kicking balls to children and adults (e.g., baseball, soccer, kick ball, etc.). They could be used to assist in kite flying, frisbee throwing, and other related recreational activities.

In general, since sports and recreation are viewed as people-oriented activities, it may take considerable time to accept and adjust to perhaps having a robotic team player or opponent.

Engineering in the Robotics Age

As mentioned earlier, robots will probably have the biggest impact in 2010 in engineering and manufacturing applications. Industrial robots used in heavy manufacturing, such as automotive assembly and painting, are commonplace. Robots are also used in construction, bridge building and repair, sanitary clean-up (e.g., oil spill clean-up), and other related tasks. Engineering "design" hasn't utilized robotics as widely as engineering "implementation." Certainly, computers and automated decision support tools are actively used in engineering design (such as developing blueprints, flowcharts, etc.), but robots in 2010 are applied to mainly engineering "construction/implementation" instead of engineering "design."

Most applications of engineering — civil, mechanical, aeronautical, environmental, electrical, industrial, etc. — are using robots in one form or another. Robots are being used in precise, delicate tasks such as integrated chip manufacturing, and robots are being used for larger, pick and place applications too (such as construction, loading/unloading, etc.).

Multiple, cooperating robots are also an essential part of the assembly line. In the future, the integration and interaction of these robots will continue to improve.

The Environment and Agriculture in the Robotics Age

In 2010, robots are commonly used in environmental disasters (such as oil spill clean-ups), space and water exploration, planting and cutting of trees, and other environment-related tasks. Robots in 2010 are biodegradable, decomposable after performing their jobs. Robots are utilized heavily in recycling plants, and even used for trash disposal and collection. They are used in scientific research such as aiding in medical experiments for contagious diseases.

Robotic machinery is also employed in agriculture. Crop harvesting, planting, milk production and pasteurization, and herding of animals are examples of where robotics are used. Robots for agricultural applications are utilized in developed nations, but have greater applicability in developing countries. Robot prices have dropped, but are still relatively expensive for the developing nations. In addition, there is plenty of cheap labor available in the developing nations to perform the agricultural tasks that robots are doing.

Education in the Robotics Age

Education in 2010 identifies the "teacher" as the facilitator or coach, as opposed to the owner of knowledge. The teacher will give suggestions to the students for navigating and searching for pertinent information. The teacher will encourage "active" learning, will have "virtual" classrooms, and will easily communicate electronically with students from different regions and countries.

How do robots fit into the educational scene in 2010? Teachers certainly have not been replaced by robots, although robots are used in the classroom as a *resource,* since they contain a wealth of information that can be accessed by the students. Robots are also being used to perform some janitorial services for the school, such as sweeping and mopping the floors in the corridors. Robots are used as aides to help the teachers in suggesting and developing lesson plans, and to allow the teacher to tap into the robot's knowledge base for obtaining information for class. The use of desktop videoconferencing, distance learning, virtual classrooms, and the "knowledge highway" is more pervasive in education than the use of robots.

Employment in the Robotics Age

With the new age of robots, the workforce now has mechanical friends as part of the office team environment. As previously said, the industrial, heavy manufacturing factory floor is where most of the industrial robots have joined the human workforce. As such, concerns by labor unions on the proportion of robots used in the total workforce have real implications. Human jobs are being replaced by many robots on the assembly floor. These jobs that are lost have created extensive company retraining, retooling programs to transition those who lost jobs (due to robotics) into new positions. New jobs have also been created due to robots, such as positions for repairing the robots, robotic engineers, and artificial intelligence specialists.

A question that has surfaced is the longevity and durability of a typical robot. With proper preventive maintenance and with advances in technology, the typical 2010 robot will last 3 years. Newer techniques and improved technology, such as in computer vision and sensing, will continue to push the state of the art of robotics.

Transportation in the Robotics Age

By 2010, cars in the U.S. will be equipped, as a normal feature, with an on-board intelligent vehicle transportation system. This system will dynamically assess the road and traffic conditions as the car is traveling, and will suggest the "best" or "alternative" routes to take to get to the destination pint. They will also provide information on best routes for trips, obtain weather information through integration to geographic information systems, and provide three-dimensional, virtual reality displays of maps for driving certain routes.

Robots are being used in the transportation field for directing traffic at malfunctioning lights and intersections, helping in building cars, trains, jets, ships, and other transportation vehicles, and assisting in road construction and maintenance. Robots are also being utilized as test pilots, test "dummies," and test clones. In the future, safety standards and laws will be invoked that relate to the use of robotics, especially in the transportation field.

Beyond the Robotics Age

It's the year 2015. Jason is 28 and he has his doctorate. Kenny is 26 and in graduate school. With the money that Kenny has saved, he is looking forward to buying a pet shop. Mom is 30 (well, at least that's what she says!). Dad still writes books, but he no longer needs to type like the old days. He can talk to the computer, and the computer will encode his words and thoughts. Dad sometimes gets into his books too much! But everything has to change sometime; however, one thing will never change and that's the love in Jason and Kenny's family.

All in all, I'd rather be at the beach!

Index